Technologies of Magic

Technologies
of Magic

A CULTURAL STUDY OF GHOSTS, MACHINES AND THE UNCANNY

Edited by John Potts and Edward Scheer

POWER PUBLICATIONS

Published by
Power Publications
Power Institute Foundation for Art and Visual Culture
University of Sydney
NSW 2006 Australia
www.arts.usyd.edu.au/departs/arthistory/power

Managing Editor	Victoria Dawson
Assistant Editor	Kirsten Krauth
Consulting Editor	Julian Pefanis
Executive Editor	Roger Benjamin
Cover design	Victoria Dawson, photograph by Heidrun Löhr
Internal design	Marian Kyte
Printed by	Southwood Press

National Library of Australia Cataloguing-in-Publication data:

Technologies of magic : a cultural study of ghosts,
machines and the uncanny.

Includes index.
ISBN 9780909952358.

ISBN 0 909952 35 3.

1. Technology - Social aspects. 2. Technology - Religious
aspects. 3. Religion and culture. 4. Supernatural.
5. Magic. I. Potts, John, 1959- . II. Scheer, Edward.

303.483

[CONTENTS]

[JOHN POTTS AND EDWARD SCHEER]

Introduction

Magic too is a myth but myths shape our machines into meanings
(Erik Davis 1999: 189).

Why is it that many technologies, particularly media technologies, continue to be shrouded in a mystique, preserving forms of magical belief within rationally ordered societies? The Enlightenment of eighteenth-century Europe, founded on principles of rational enquiry, was supposed to have expunged the mystical from thought and from cultural expression. Yet mystical belief has persisted, even attaching itself to succeeding generations of technology. Nineteenth-century spiritualism was enchanted by the uncanny doubles generated by photography and cinema; techno-mystics of the early twentieth century probed the spirit world through the means of radio. In the twenty-first century, mysticism has moved, like everything else, into cyberculture. The uncanny extension of biological modes of perception by technological means continues with virtual imaging, avatars, and the dreams of "uploading" the self into an immaterial state.

The territory charted by the essays in this book is the curious field occupied by both machines and magic. It is the techno-mystical domain of science and the spirit, technologies and ghosts. Refracted through highly technologised, post-Enlightenment societies, magic manifests itself in surprising ways and through an enormously diverse range of practices, as the essays in this volume suggest. Yet they all investigate, in various ways, the co-existence of very old forms of thought—belief in magic, ghosts, spirits—and contemporary practices. Although employing different approaches to the topic of magic—from magic as metaphor to magic as ritual and performative mode—the writers represented in *Technologies of Magic* resist a strict dichotomy between the age of scientific reason and the "primitive" belief in the powers of magic.

THE DISCOURSES OF MAGIC

Such a dichotomy was advocated by earlier writers such as J. G. Frazer who disparaged magical practices as flawed logic, inferior forerunners of scientific method. In *The Golden Bough: A Study in Magic and Religion* (1922), Frazer described magic as "a spurious system of natural law as well as a fallacious guide of conduct" (1974: 15). Frazer installed a discontinuity between himself—as anthropologist and man of reason—and the misconceptions of the "primitive" mind; it was the role of the enlightened thinker to illuminate the murk of the past, to "discern the spurious science behind the bastard art" (Frazer 1974: 15).

Later theorists have criticised Frazer's approach to the study of magical systems. The philosopher Ludwig Wittgenstein, writing in 1931, found Frazer's condescending interpretation "wrong-headed". Wittgenstein opposed the strict division between the past (primitive, misguided) and the present (rational, enlightened). For Wittgenstein, there is a continuity of thought throughout history and across cultures about the idea of magic, which is "much more general than Frazer shows it to be" (1993: 5e). The principle of magic is readily observed in many contemporary practices, from superstitions to private rituals. Whereas Frazer insisted that magical practice is fundamentally alien to modern thought, Wittgenstein proposed that it is perfectly possible for us to think in magical terms. We could, for example, imagine that "in a given tribe no-one is allowed to see the king, or again that every man in the tribe is obliged to see him" (1993: 5e)—an instance of magical thinking.

Indeed, for Wittgenstein, the continuity of magical belief is guaranteed by the very languages we speak. When Frazer offers an interpretation of a tribal observance based on the irrational fear of a ghost, Wittgenstein remarks that the word "ghost" entails that Frazer "understands this superstition well enough, since he has used a familiar superstitious word to describe it" (1993: 8e). We have a "kinship" with tribal peoples through language; words such as "ghost" express "something in us too that speaks in support of those [tribal] observances" (1993: 8e). For Wittgenstein, the fact that "we include the words *soul* and *spirit* in our civilised vocabulary" (1993: 8e) ensures the persistence of magical belief: a "whole mythology is deposited in our language" (1993: 10e).

Recent theoretical approaches in cultural history, anthropology, and technology studies have favoured Wittgenstein's perspective, as a brief sketch of the field demonstrates. Bruno Latour has reassessed the conceptual separation of "pre-modern" from "modern" which has performed a decisive role in separating science from magic in Western culture. For Latour and other historians of science such as Michel Serres, any historical period comprises "multiple times" (Latour 1993: 74), while even the scientific practices constitutive of modernity conjure an "incongruous blend of hybrids" (1993: 142). Serres has rejected the notion of epistemological breaks in the history of thought and culture, advocating a "technique of rapprochement" with pre-modern modes: tradition, he argues, "often gives us ideas still filled with vitality" (1995: 48, 50).

Other recent studies have revealed previously unsuspected continuities of thought across historical spans and cultural differences. In his book *The Religion of Technology* (1997), David Noble traces the intertwined paths of technology and religion from the Middle Ages to the present, asserting that "the technological enterprise…remains suffused with religious belief" (1995: 5). The "technological sublime" in Noble's account is age-old spiritual transcendentalism breathing through machines. Margaret Wertheim covers a similar historical trajectory in *The Pearly Gates of Cyberspace* (1999), travelling from Renaissance soul-space and celestial space to the spiritual yearnings within cyberspace. Erik Davis charts an even longer time-span in *TechGnosis* (1998), tracing the powerful mystical impulse from ancient gnosticism to the utopian fantasies of the internet. For Davis, "technologies can serve as the vehicles for spells, ghosts and animist intuition"; they can also "provide launching pads for transcendence, for the disembodied flights of gnosis" (1998: 6).

Cultural histories such as these constitute some of the context for the essays contained in this book. *Technologies of Magic* occupies a cross-disciplinary area, drawing on other recent work in anthropology, media studies and sociology.

Jeffrey Sconce's *Haunted Media* (2000) investigates the mysticism infusing media technologies from telegraphy to television. In *Technology as Magic* (1999), Richard Stivers finds in the contemporary media landscape "the triumph of the irrational", as advertising and the mass media sell us dreams of technology wrapped in a magical cocoon. Avery Gordon's *Ghostly Matters* (1997) is an inter-disciplinary work, drawing on, at least, sociology, literary criticism and cultural studies. Constructing a new theoretical space as she goes, Gordon brings to light the socially marginalised, the excluded and the repressed, those figures haunting the contemporary West.

Finally, Michael Taussig's work, as represented by *Mimesis and Alterity* (1993), forges new links between disciplines and writing practices. Barriers between self and other, civilised and primitive, are broken down as sympathetic magic escapes its ethnographic bonds. There is a renewed focus on magical properties, commodities and technologies as anthropology converges with other disciplines including philosophy and cultural studies. This interzonal space—a cultural anthropology of human and post-human performances, signs and objects—takes up a large part of the territory explored in *Technologies of Magic*.

DEFINITIONS OF MAGIC

This brief map of the theoretical terrain indicates the fluidity of recent approaches within technology studies, cultural history and the critical study of magic and other social practices. The definitions of magic, religion and other systems of thought and practice have themselves been transformed in recent decades. The generation of anthropologists influenced by Frazer could comfortably distinguish between magic—as a set of practices aimed at controlling nature—and religion which entailed, according to Bronislaw Malinowski (1992: 18), "propitiation of superior powers". Writing in 1948, Malinowski further contrasted magic's specialist nature, concentrated in the hands of individual practitioners (wizards, witches), with the universal participation in religion, "in which everyone takes an active and equivalent part" in prayers and rituals (88–9). Keith Thomas (1971) summarises the earlier distinction between magic and religion: magic "postulated occult forces of nature which the magician learnt to control", while religion "assumed the direction of the world by a conscious agent who could only be deflected from his purpose by prayer and supplication" (1971: 41).

As Thomas remarks, this dichotomy has been rejected by later theorists, on the ground that it "fails to consider the role which the spirits can play in a magician's ritual and which magic has occupied in some forms of primitive religion" (1971: 41). The belief in transubstantiation in the Catholic mass, that the communion recipient taking the wafer and wine is actually incorporating the flesh and blood of Christ, would be an example of a contemporary religious practice predicated on a magical premise. In fact religious rituals of all kinds often involve magical transformations such that the line between magic and religion, so clearly defined even in the early twentieth century, has become blurred.

As Simon Price and Emily Kearns observe, "magic" has always been an ambiguous term. It derives from the ancient Greek term referring to the rites of the Persian magi; this term had ambivalent usage for the Greeks. It could carry a positive connotation when applied to sages such as Zoroaster, but it could also refer disparagingly to sorcerers, frauds and quacks. By late antiquity, prayers, wizardry, philosophy, magical ritual and formulae intermingled so widely as to "challenge modern distinctions between magic and religion (and science)" (Price and Kearns 2003: 327). Price and Kearns propose a broad definition of magic to encompass these diverse strands: "a manipulative strategy to influence the course of nature by supernatural ('occult') means". This general definition incorporates an overlap with religion ("supernatural means"), while preserving—through reference to "manipulative (coercive and performative) strategy"—a distinction from religion.

Definitions of magic vary in these essays, but the key notion is the kind of performativity which arises from an agent of transformation whose effects are evident but whose operations are not apparent. Magic is therefore considered variously as a performance genre associated with illusionism, to magic as a broader cultural function which becomes visible within ritual contexts. As Michael Taussig notes, it is in ritual that "the symbolic character of the symbol erases itself" (Taussig 1999: 137) and signs revert to forces. These forces can act on bodies and transform them. Language itself can function in this manner, as J. L. Austin observed in his elaboration of the notion of the performative within speech act theory. The performative designates a class of language which acts directly on or transforms the world, such as acts of parliament or marriage vows, or the sentencing words used by a judge. All of these change the state of affairs from the moment of their meaningful utterance in a ritual context, and it is these ritual contexts which give the words their power. The performative draws upon this power to be effective. The discourse of magic spells relies on this power of language not only to describe but to enact material changes and—as Chris Chesher argues in his essay on invocation—it is also the language of computer users. The essays in this collection all examine such forces and effects in relation to technologies whose meanings and uses make visible the very mystical impulses that they were supposed to replace.

THE TECHNOLOGICAL UNCANNY

The essays in this collection are similarly concerned with the process—historical, perceptual, cognitive, metaphorical, performative—in which the lines between technology and spirit, rationality and mysticism, become fuzzy when scrutinised by the contemporary critical gaze. The focus throughout this volume, expressed through various methods, is how this strange borderland is made evident through the uncanny liveness of our machines.

When machines seem to take on a life of their own we discover a fundamental trace of the uncanny. It was E. T. A Hoffmann's automaton in his story "The Sandman" that first brought this dimension of experience to the psychologist Ernst Jentsch, who constructed of it a quintessential instance of the uncanny. Freud quotes Jentsch at the start of his own inquiry into the uncanny, in his essay "Das Unheimliche" from 1919. Freud cites "'doubts whether an apparently animate being is really alive; or conversely, whether a lifeless object might not be in fact animate'" as Jentsch's key example, referring to "the impression made by waxwork figures, ingeniously constructed dolls and automata" (Freud 219). The idea is subsequently developed by Freud to indicate the appearance of the strange within the familiar. Freud argued in an extensive etymology of the terms "*heimlich*" (homely, familiar) *and* "*unheimlich*" (eerie, strange) that the most frightening experiences we can have are based on those moments of uncertainty when the familiar has been made strange and secret through repression, but which then returns.

Jentsch proposed that, in literature, "one of the most successful devices for easily creating uncanny effects is to leave the reader in uncertainty whether a particular figure in the story is a human being or an automaton." (Freud 219) We wish to revisit this idea here. In one sense it is a proto-cybernetic perception not simply of the familiar human element revealed in the automaton, but of that moment when the human/machine interface disappears and the system behaves with characteristics independent of either category. This for Jentsch unleashes the radical ontological uncertainty of the uncanny.

Whereas Freud focused on "the peculiar emotional effect of the thing", this collection of writings negotiates the power and the contemporaneity of this experience. One of the features of the essays collected here is the way in which they configure the strange relations between machines and humans who, despite everything, still do not understand each other very well, as powerful transmitters and transducers of the magical in contemporary culture.

This question of the uncanny in relation to particular technological formations is perhaps still recuperable as an instance of the Freudian uncanny, that is, as the return of the repressed. As the essays in this collection show, the acceleration of technological development runs alongside an equal and opposite insistence on the far side of the technological: the irrational, the unconscious. It is as if the more stuff we have to use and the more there is to explain, the less satisfied we are with the explanations given by the producers of technology. There is a sense, perhaps post-genome that, as more and more of what was once considered essentially human becomes digitised and therefore alterable, hackable, the stronger is the desire for experiences which seem to resist the quotidian photoshop of digital culture. Yet the cybernetic experience is one in which human life is inextricable from the life of its technologies.

More recently in the field of technology, robotics has explicitly confronted the issue of the uncanny. A study conducted by Masahiro Mori in the late 1970s was about the pattern of human emotional response to robots across a range of anthropomorphic resemblance. Mori's controversial idea of the "uncanny valley" is taken from the graphic representation of this response which indicates that as a robot is made "more humanlike in its appearance and motion", the corresponding emotional response from a human being to the robot becomes "increasingly positive and empathic until a point is reached at which the response suddenly becomes strongly repulsive". The graph showing emotional response against anthropomorphism falls rapidly away, before climbing again as the robot's appearance and motion "are made to be indistinguishable to that of human being", creating the valley effect of the name.[1]

The machines which populate this collection are all particular artefacts of technology or manufactured artificial mechanisms: cameras, electric lights, robots—but not all are residents of the "uncanny valley". Yet they all reveal, especially in their moment of emergence, the way that technological development produces an emotional and ideational supplement beyond the immediate uses and purposes of the technology. This excess is a necessary by-product of the interaction between organic and non-organic life and is, in turn, managed by other forms of machine, that of the state or of industry such as the cultural sector or the computer industry.

THE STRUCTURE OF THIS BOOK

The fundamental connections linking the appearance of new machines and the colourful excess of human responses are made evident in slightly different ways across the three parts of this book, namely: "The Persistence of Magic in Modernity", "Ghosts and Their Machines" and "New Technologies and Their Doubles". Part One is concerned with the feedback loop by which the archaic disrupts the modern. Scott McQuire's essay explores "The Persistence of Magic in Modernity" through a historical study of the development of a technological uncanny. McQuire charts a cultural history of the mystique surrounding the appearance of electric light in urban space. He describes the "widespread transference of the romantic experience of nature onto technology at the end of the nineteenth century". McQuire writes of the emergence of terms such as "live wires, human dynamos and electrifying performances" at the same time as electricity "entered daily life in the 1880s". In this essay the magical property of electricity is not only its phenomenal luminescence but the inspiration it provided: "(t)o feel electricity in the air became synonymous with excitement, arousal and even love".

Rachel Moore's essay "Love Machines" considers the ways in which old movies, themselves an outmoded technology, present the magical (transformative, performative, miraculous) moment of falling in love. This is achieved through the intervention of various outmoded technologies, as if to argue for the cinema itself as a magical technology. It is in this kind of cinema, Moore argues, that "the mythic (in this case the primal passion of erotic love) travels from prehistory to the present, thus providing a magical modern conduit for the entry of the archaic into the modern."

The other essays in this part develop this theme in their own ways. Annette Hamilton's study builds on this topic of object-focused affect through an anthropological exploration of alternative conceptualisations of objectivity and "things". Her essay points out that these things no longer live their lives wholly distinct from the world of human affects. In some cases the object's sole purpose is to become a kind of emotion-generating machine. Here the technological object has become the psychoanalytic part-object, an intimate object split off from the maternal body. In this way it has moved into the uncanny territory of both Jentsch and Freud.

The fourth essay in this part of the book deals with stage magic—or illusionism—rather than ritualised magic. Patricia Pringle discusses interior design as a "practice of natural magic" which Pringle defines as "an attempt to arrange life for maximum emotional and practical power". Pringle locates this kind of practice in nineteenth-century stage illusionism. She draws on examples from stage illusions to discuss the "human fascination with 'impossible' operations, such as dematerialising, defying gravity, vanishing, shimmering, hiding, folding up, or changing form". Her argument looks at how these magical/performative fascinations manifest in that other occult practice of interior design.

Part Two, "Ghosts and Their Machines", takes as its point of departure the notion that the cybernetic experience can be seen as both distinctly contemporary and, equally, redolent of the past. There is an explicit reminder of the "primitive" spiritualism of previous generations of photography and moving images in which ghosts were conjured from the chemical manufacture of likenesses. But are these other post-human presences of the same order as the chemical and electromagnetic ghosts of previous technological ages? Does the nature of ghosts and magic change with the technologies to which they are linked? In any case how do we negotiate territory with them? The essays in Part Two constitute one such negotiation, or at least an attempt to engage with these questions.

The figure of the ghost in the machine operates as a critique of the Cartesian model of mind separated from body; yet—as Anne Cranny-Francis argues—it is also at large in the worlds of cognitive science and artificial intelligence (AI) where the machinery of brain or computer never quite matches the ghost of consciousness. Yet this kind of science is haunted by the apparitions it sees but disallows. Cranny-Francis' essay takes us back to the frontier of Enlightenment politics where experience which cannot be repeated can only be disavowed.

The subject of John Potts' essay is the ghost as an idea that recurs across history and across cultures. Potts proposes that the core of this idea is the representation of the past: the ghost is an immaterial presence haunting the living, but it is also the past haunting the present. Thus haunted houses preserve the memory of former residents; "hungry ghosts" maintain the memory of ancestors. Potts also suggests that this core idea is transformed in accord with its cultural and historical context. The ghost is an idea that performs cultural work, meeting the demands of specific societies in which it functions. Potts analyses the contemporary popular discourse on ghosts as evident on the WorldWideWeb: he finds that this particular inflection of the "ghost-idea" expresses a contemporary need to filter mystical belief through a post-Enlightenment framework, including classification systems and pseudo-scientific method.

In a different part of the landscape, the ghost functions as a metaphor in post-colonial studies, as the trace of repressed voices now re-emerging from beneath dominant discourses such as in Stephen Muecke's report from Madagascar. This chapter also wants to account for what is traditionally excluded from scientific discourse—here the discourse of the human sciences, and specifically anthropology. Muecke points out that "Island of Ghosts" is famous for its primates that carry its older name of Lemuria. Lemures, in Roman religion, were "vampire-like ghosts of the dead." He looks at rituals about these for an aesthetic of contingency and chance, as a way of maintaining an open dialogue with the culture in its own terms. These terms give ghosts and magic a place, before the application of an ethnographic terminology which would render these entities as simply social or pathological phenomena. His paper examines the ways in which this aesthetic of contingency informs current developments in performance studies and cultural studies by interrogating some of the repressed content of anthropology. Muecke's open approach can also deal with the troublesome machinery of ethnographic research, principally its recording devices, which are not entirely neutral and produce their own ghostly emissions. It is as he says, "the unreliability of the machine which can be productive of the ghostly presences, and so too can its cultural excesses: William Ellis's photography in Madagascar, far from being a reliable witness, was a miracle-machine in a political contest for another intangible, Christian faith."

Part Three of the book, "New Technologies and Their Doubles", consciously invokes Antonin Artaud's ghost and that of Freud to examine the technological uncanny from the perspective of ritual. There is a focus on the performative dimension in ritual and culture, the dimension of forces and transformations—an Artaudian dimension. It also broaches topics as diverse as the notion of the mechanistic cosmos to theories of extended mind, emergence and cybernetics. It returns us, strangely, inevitably, to Freud's uncanny essay in which he theorises the double in the following terms:

> For the "double" was originally an insurance against the destruction of the ego, an "energetic denial of the power of death", as Rank says; and probably the "immortal" soul was the first "double" of the body.... Such ideas... have sprung from the soil of unbounded self-love, from the primary narcissism which dominates the mind of the child and of primitive man. But when this stage has been surmounted, the "double" reverses its aspect. From having been an assurance of immortality, it becomes the uncanny harbinger of death...The "double" has become a thing of terror, just as, after the collapse of their religion the gods turned into demons (219–20).

The notion that technology itself would double human achievements and behaviours, in both the utopian sense and the monstrous sense of the double, is still with us as research into prosthetics, robotics and AI indicates. But technology generates its own doubles and produces its own phantoms. Andrew Murphie's investigation into recent theories of the brain reveals contemporary "cultures of cognition", ways of thinking about the brain which construct it as an uncanny double for technology, the magical machine room of consciousness. Murphie considers magic as a hidden causal agent, as "the force of transformation, as active participation in the unknown (if not immaterial), or as the intensity of social actions that mediate different aspects of the material world". In these terms the brain becomes "perhaps *the very* figure—of magical transformations of forces" because, as he argues, while the brain is a "component of the body's exchange of codes and energies" it does not itself signify. Its own functions and operations are secret even while its effects are everywhere in evidence.

Murphie's use of magic as a cultural performance, a kind of ritual, in which "forceful acts of transformation" occur, is echoed in Chris Chesher's discussion of the centrality of a kind of performative language within these rituals of transformation. Chesher's research into the importance of invocation in computer user interfaces is carried out in the form of a cultural history or a conceptual biography which examines the life of invocation from Greek mythology to current digital interfaces. It underscores the links between this everyday use of technology and a variety of ritual practices in which language functions in this way. In this sense invocation is a similar notion to Austin's performative speech act, in that the latter activates transformations in the material world through repetition in ritual contexts. Similarly Chesher focuses on the citational quality of invocation "embedded in technological systems". It is less specific than the performative and more ubiquitous, since it is "embodied in the design of computers" themselves. It is finally a linguistic entity which straddles technological practices and ritual/magical ones.

Edward Scheer's discussion of the performance art of Stelarc as "cyber-ritual" also develops the technological uncanny in terms of the doubles generated by technology. Scheer investigates a specific Stelarc performance called *Movatar* (2000), in which a specially designed avatar was connected to Stelarc's body through a hydraulic actuator. In this piece Stelarc plays quite purposefully with the idea of the uncanny double by constructing a system in which his body functions as the real world "prosthesis" for the avatar. Scheer reads this work in terms of sympathetic magic and Stelarc's work in general as an example of a contemporary "life crisis ritual": it is "the life crisis of an obsolete body finding itself without sympathetic environments in an age of technological innovation which is accelerating beyond the capacities of the organism to adapt". Scheer argues that Stelarc's cyber-rituals of the end of independent human evolution imagine and perform the extension of the organic body's capacities for action in a way which suggests "that the idea of a 'natural body' disconnected from technology is itself a phantom, a ghost of the Enlightenment".

As Erik Davis argues, it is the magician's task to manipulate creatively the phantasms of the social: the "images, stories and desires" through which the world makes itself apparent. Scheer argues that "these phantasms as much as the tired old human organism itself, are the stuff, the material of Stelarc's artwork". Stelarc's art of uncanny machine-life makes him a highly visible model for the contemporary form of magic this book has begun to examine.

This suggests a way of understanding the distinctive feature of this collection, since the ghosts and the magic it examines are framed as fundamentally performative, in the sense of performing new kinds of work in the world. The ideational excess accompanying technological developments is not considered here as waste or delusion, but as productive of new modes of social and cultural engagement with technology. The ghosts of techno-culture are not asked to identify themselves or show their credentials, but are asked instead to account for the work they clearly do. These essays also show how the apprehension of the magical in the world of machines may give rise to a feeling of uncanny unease, but that this ultimately produces another set of relations between organic and non-organic life, and another way of seeing the place of technology in the cultural world.

ENDNOTE

1. http://en.wikipedia.org/wiki/Robotics

REFERENCES

Davis, Erik (1999) *TechGnosis: Myth, Magic + Mysticism In the Age of Information* London: Serpent's Tail.

Frazer, J. G. (1974) [1922] *The Golden Bough: A Study in Magic and Religion* London: Macmillan.

Freud, Sigmund (1953) "The Uncanny," in *The Standard Edition of the Complete Psychological Works of Sigmund Freud* (ed. and transl. James Strachey) XVII London: Hogarth 219–52.

Gordon, Avery F. (1997) *Ghostly Matters: Haunting and the Sociological Imagination* Minneapolis: University of Minnesota Press.

Latour, Bruno (1993) *We Have Never Been Modern* (transl. Catherine Porter) New York: Harvester Wheatsheaf.

Malinowski, Bronislaw (1992) [1948] *Magic, Science and Religion and Other Essays* Illinois: Waveland Press.

Noble, David F. (1999) *The Religion of Technology: The Divinity of Man and the Spirit of Invention* New York: Penguin.

Price, Simon and Kearns, Emily (eds.) (2003) *The Oxford Dictionary of Classical Myth and Religion* Oxford: Oxford University Press.

Sconce, Jeffrey (2000) *Haunted Media: Electronic Presence from Telegraphy to Television* Durham: Duke University Press.

Serres, Michel with Latour, Bruno (1995) *Conversations on Science, Culture and Time* (transl. Roxanne Lapidus) Ann Arbor: University of Chicago Press.

Stivers, Richard (1999) *Technology as Magic: The Triumph of the Irrational* New York: Continuum.

Taussig, Michael (1993) *Mimesis and Alterity: A Particular History of the Senses* New York: Routledge.

------------(1999) *Defacement: Public Secrecy and the Labour of the Negative* Stanford: Stanford University Press.

Thomas, Keith (1971) *Religion and the Decline of Magic* London: Weidenfeld & Nicolson.

Wertheim, Margaret (1999) *The Pearly Gates of Cyberspace: A History of Space From Dante to the Internet* Sydney: Doubleday.

http://en.wikipedia.org/wiki/Robotics accessed 20 May 2005.

Wittgenstein, Ludwig (1993) *Remarks on Frazer's* Golden Bough (transl. A. C. Miles) Denton: The Brynmill Press.

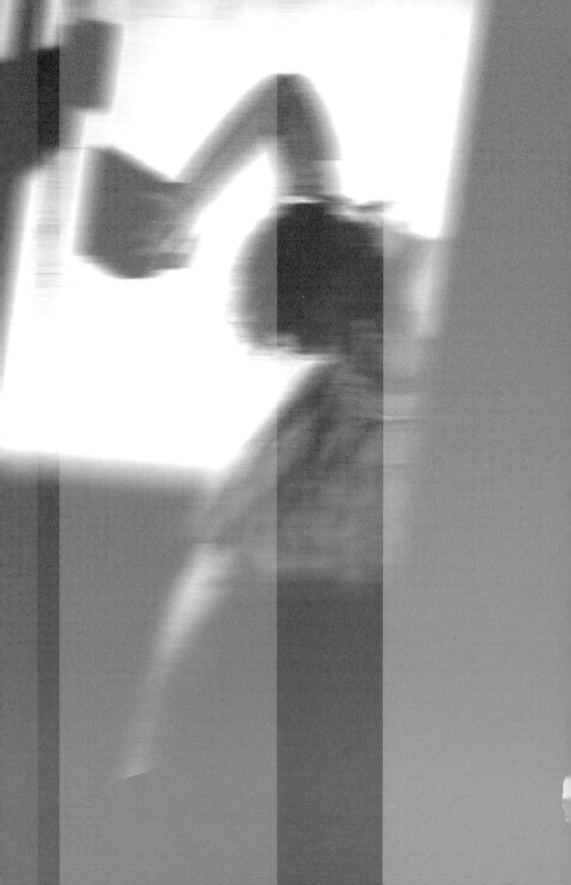

part ONE

[THE PERSISTENCE OF **MAGIC** IN MODERNITY]

[SCOTT McQUIRE]

Dream cities:
The uncanny
powers of
electric light

Electricity is the pervading element that accompanies all material existence, even the atmospheric. It is to be thought of unabashedly as the soul of the world (Goethe 1825).

If you build buildings with lights outside, you can make them indefinite, and then when you're through with using them you shut the lights off and they disappear (Andy Warhol 1975).

In his famous 1919 essay, Freud (1955: 219–52) defines the uncanny to include experiences in which inanimate objects seem to come to life. In early modernity, this sense of the uncanny accompanied the spread of electric light, itself a manifestation of the near-miraculous powers of electricity. From the moment of its initial recognition as an independent phenomenon, electricity has been a source of profound wonder. Romantics rapidly identified it with a universal life force, dramatised in the archetypal modern creation scene of Mary Shelley's 1818 novel and distilled by Goethe into "the soul of the world". A century later, the prospect of widespread electrification literally dazzled the world, inspiring entrepreneurs, artists and revolutionaries alike with visions of an irresistible electrical future.

At the same time, electricity has always led a double life. Beneath the Promethean narrative of limitless possibilities lies a more utilitarian tale of practical development. Counterpointing the arcane myth of electricity's magical properties—force without muscle or steam, light without flame—is the profane physical reality of its often cumbersome technical infrastructure. Supporting the spark of the incandescent lamp which shines brighter than any jewel are unsightly poles and criss-crossing wires, not to mention ferocious patent wars and internecine struggles to form some of industrial capitalism's most powerful corporations.

This split identity is replicated in the literature about electricity. While there is a plethora of accounts of inventor-engineers such as Edison or Tesla, or of giant corporations such as General Electric and Westinghouse which grew from local and regional businesses to exploit the new technology, there is a relative dearth of social histories examining the difference electricity made to everyday life.[1] Even scarcer are accounts of the way in which electricity has contributed to the formation of a distinctively modern sense of space, most dramatically through the electrical illumination of the modern cityscape which has so decisively changed experience of the urban environment.

In the absence of systematic accounts, frequent snippets can be found scattered through the writings of artists, architects, journalists, filmmakers and other observers of the modern city. One thing most of these reports make clear is that, even from the first, electric illumination exceeded a purely functional role. This excess over and above any pure functionality was already apparent in the earlier generation of public experiments with electricity conducted by those such as Sturgeon and Saxon in London in the 1830s. Morus (1998) has emphasised the spectacular nature of their displays, concluding that the creation of special effects such as giant sparks was an indispensable element not only in attracting public attention but in winning the battle of public opinion.

In 1885, when Edison's incandescent lamp was less than a decade old and the illumination of public space a rarity, a scheme was mooted for lighting the entire city of Paris with what was grandly dubbed an artificial sun. The plan comprised one hundred 200 000 candlepower lamps mounted on a single tower soaring 1100 feet (335 metres) in the Tuilleries Gardens. The fact that the scheme was both impossible, because lamps of such magnitude had not yet been invented, and impractical because it would only light the cityscape from one side, merely underlines the extent to which the very idea of electrical illumination has long had a powerful symbolic role. By the 1880s, when electrical systems began to be widely adapted to practical uses, electricity was seen as the key to achieving a new level of control over the lived environment. The ability to convert night into day at the flick of a switch offered the most striking proof of the superiority of the modern present over the past, the most compelling evidence of the ability of technological progress to subdue nature.

Equally telling is the rapture with which many people greeted their first sight of electric light. Only four months after Edison's famous demonstration at Menlo Park, the city fathers at Wabash hired the Brush Co. to set up four 3000 candlepower arc lights on the courthouse. The event attracted 10 000 visitors to the small town. The local press reported:

> People stood overwhelmed with awe, as if in the presence of the supernatural. The strange weird light exceeded in power only by the sun, rendered the square as light as midday ... Men fell on their knees, groans were uttered at the sight, and many were dumb with amazement. (Nye 1990: 3)

It is probably wise to take such a tale with a grain of salt—after all, boosterism is grist to the mills of small-town papers, which also reported that local farmers could expect giant pumpkins and corn stalks as a result of the new light. Yet, the spectators' reaction should not be discounted too quickly. Even read as an apocryphal tale in the genre of credulous cinema audiences fleeing the image of an onrushing train, it registers the extent to which electricity departed all previous protocols of illumination. Before the 1880s, artificial light came only from fire. Candles, kerosene and even gas were smoky, potentially dangerous flames whose ability to illuminate was clearly linked to their consumption of fuel. By contrast the enclosed, vacuum-sealed light bulb, was a paradox, producing a light which was intense, yet smokeless, fireless and seemingly inexhaustible. Electric light produced a categorical anomaly, one which was experienced by nineteenth-century observers not so much as monstrous but miraculous.

The image of the Wabash public gazing at arc lights in silent awe indicates the extent to which electrical illumination belongs to what Leo Marx (1965) has aptly called the technological sublime. For Marx, the phrase describes the widespread transference of the romantic experience of nature onto technology at the end of the nineteenth century. One of the key sites for this transference was the appearance of massive industrial machinery such as the electrical turbine, an apparatus which generated not only electrical current but an irresistible series of concepts and metaphors. As electricity entered daily life in the 1880s, live wires, human dynamos and electrifying performances all became recognisable descriptors for a specific form of modern energy. To feel electricity in the air became synonymous with excitement, arousal and even love.

It was in this context that electricity spread through the modern cityscape in several waves. Initially confined to the mansions of the wealthy and a few department stores seeking a novel means of attracting shoppers, it expanded into public street lighting schemes along major transport routes, before finally extending into most private homes. Electrification of the home altered domestic space rapidly and significantly. Electric light was not only much brighter than candles, kerosene and even gas; it was cooler, and, importantly, less prone to fire. Because people did not have to huddle together around a dim flame, electric lamps enabled a heightened level of independent activity, contributing to a significant increase in the practice of reading. The extension of grid lighting to domestic consumers was also the conduit along which a range of electrical appliances manufactured by the big power companies such as General Electrics and Westinghouse was launched into the home: irons, toasters, heaters, cookers, refrigerators and radios were the first in what has become a seemingly inexhaustible list.

In the street outside, the changes were even more dramatic. The possibilities for using electric light to alter the appearances and ambiance of urban space were first systematically explored in the series of world fairs stretching from the 1880s to the First World War, as governments and corporations combined to put coherent visions of a fully electrified society on public display. The 1876 Philadelphia Exhibition is notable in being the last major exhibition based on steam power; it was also one of the last closed at night. After the 1879 London Exposition featured Edison's new invention as a chief attraction, subsequent fairs became key sites for lighting innovation, as each city sought to outdo its rivals in the number of lights and the power of their illumination.

The Chicago World Fair of 1894 had more lights on its Electrical Building alone than were used by the entire Paris Exhibition of 1889 for which the Eiffel Tower was built; the Chicago fairgrounds also contained more light than any contemporary city in the USA. As Nye (1990: 37) points out, millions of visitors to these fairs saw more artificial light than they had ever seen in their lives.

While world fairs and similar exhibitions were not profitable in themselves, the National Electric Light Association in the USA noted their value as "load builders" instrumental in increasing demand for street lighting and other uses of power. So profoundly depressing was the unilluminated cityscape that the imposition of blackouts in US cities during the First World War led to immediate calls for their withdrawal. Visitors to the fairs saw artificial light used in dramatic new ways: to delineate the outlines of buildings and pathways, to illuminate fountains and water jets, to probe the depths of the night sky. *Cosmopolitan*'s reporter described the scene at Chicago in what can only be called glowing terms:

> Look from a distance at night, upon the broad space it fills, and the majestic sweep of the searching lights, and it as if the earth and sky were transformed by the immeasurable wands of colossal magicians and the superb dome of the structure that is the central jewel of the display is glowing as if bound with wreaths of stars. It is electricity! When the whole casket is illuminated, the cornices of the palaces of the White City are defined with celestial fire (Nye 1990: 38).

Soon the emphasis began to move away from the sheer quantity of lights to the use of hidden lighting which enabled buildings to be displayed as striking forms in integrated artificial landscapes. These techniques migrated rapidly from the idealised urban spaces of the world fairs, into more prosaic but no less fantastic environments such as the amusement parks on Coney Island. Maxim Gorky's visit to Luna Park in 1907 found him entering a fabulous terrain composed of 1.3 million lights:

> With the advent of night a fantastic city all of fire suddenly rises from the ocean into the sky. Thousands of ruddy sparks glimmer in the darkness, limning in fine, sensitive outline on the black background of the sky shapely towers of miraculous castles, palaces and temples. ... Fabulous beyond conceiving, ineffably beautiful, is this fiery scintillation (Koolhass 1994: 29).

Other spectacular forms of illumination rapidly appeared at this time including the electrical advertising sign. The first blinking sign, spelling E-D-I-S-O-N, had been shown at the London Exhibition of 1882. By 1900, the use of commutators made it possible to organise visual sequences which could produce the illusion of motion, exploiting the same trick of persistence of vision which was used by cinema. By 1910, more than 20 blocks on Manhattan's Broadway were covered in electrical advertising. The intensity of illumination lent the thoroughfare its famous sobriquet, and the "Great White Way" was soon to be imitated by countless cities and towns making their own claim to being "modern". If such an environment offended the beaux-arts aesthetic, it was ready-made for the avant-garde. On his arrival in New York in 1917, Marcel Duchamp famously declared the entire city a work of art. When the great revolutionary poet and modernist proselytiser Vladimir Mayakovsky visited New York in 1925, he was impressed above all by the lights of Broadway:

> The street lamps, the dazzling lights of advertisements, the glow of shop windows and windows of never-closing stores, the lights illuminating huge posters, lights from the open doors of cinemas and theatres, the speeding lights of automobiles and trolley cars, the lights of the subway trains glittering under one's feet through the glass pavements, the lights of inscriptions in the sky. Brightness, brightness, brightness... (Woroszylski 1971).

One of the most elaborate advertising signs, standing 72 feet (22 metres) high and 90 feet (27 metres) wide, sat atop the Hotel Normandie creating the illusion of a 30 second Roman chariot race. The *Strand Magazine* reported:

> It is more perfect and natural in its movement than the finest coloured cinematograph (Nye 1990: 52).

This comparison alerts us to the extent to which the electrification of the modern city created a new perceptual matrix which strikingly paralleled the experience of cinema. The coincidence is still worth remarking. At the same moment that electric light charged the cityscape with spectacular effects previously reserved for specialised showplaces, the spread of new modes of rapid transit and the proliferation of glass architecture worked to set every urban traveller's eye on a collision course with this shimmering, phantom city.

This fusion of light and movement rapidly became a hallmark of the modern city, establishing a spatiality both exhilarating and potentially disorienting to its inhabitants. What emerges for the first time is an *other* city, an oneiric city which exists only at night and whose dream forms have only tenuous connections to the prosaic spaces of the waking day. In the USA large corporations began to promote themselves by floodlighting their skyscrapers. Icons of the age, beginning with the Singer Building in 1907, were baptised in light, the expense justified by their conversion into blazing symbols visible to millions. The Woolworth Building, which took over the mantle of world's tallest in 1913, was designed with its electric illumination in mind.

In France, Le Corbusier's characteristic enthusiasm for new technology emphasised the possibilities for the electrical transformation of architecture:

> One Armistice Day in the evening, M. Citroën offered us that undreamed of revelation: a floodlit Place de la Concorde. Not just lit up by its street lamps, or the Republic's standardized little gas flames, but illuminated with all the floods of light made possible by electricity. The idea had come from America, the projectors from the war. It was (and continued to be every evening) one of the most astounding lectures on architecture that it would be possible to attend "in this wide world". Sublime straight lines, and oh, sublime French rigor! On that Armistice night a dumbfounded crowd standing in the square, held in the grip of a grace unshadowed by a single jest—on the contrary, of a grace imperious in its command—that crowd was able to listen *to architecture itself* (Corbusier 1964: 178).

Corbusier's vision of electric light converting mute architecture into a living, *speaking* entity situates the uncanny nature of the new technological landscape. It also indicates an ambivalence which dogs the electrified city, undermining every attempt to split Corbusier's radiant city from its supposedly irrational double, the nightmare city at the dark heart of expressionism and film *noir*. In fact, the city of light and the city of night are recto and verso of the same developmental forces, but the dream of their bifurcation structured both the key theoretical treatises of modern architecture, and a host of popular narratives. Exemplary of the latter is Thea von Harbou's novel *Metropolis* (which formed the basis for husband Fritz Lang's epic film in 1926):

> The workman No. 11811, the man who lived in a prison-like house, under the underground railway of Metropolis, who knew no other way than that from the hole in which he slept to the machine and from the machine back to the hole— this man saw, for the first time in his life, the wonder of the world, which was Metropolis: the city, by night shining under millions and millions of lights.
>
> He saw the ocean of light which filled the endless trails of streets with a silver, flashing luster. He saw the will-o'-the-wisp sparkle of the electric advertisements,

lavishing themselves inexhaustibly in an ecstasy of brightness. He saw towers projecting, built up of blocks of light, feeling himself seized, over-powered to a state of complete impotence by this intoxication of light, feeling this sparkling ocean with its hundreds and thousands of spraying waves, to reach out for him, to take the breath from his mouth, to pierce him, suffocate him ... (von Harbou: 50–1).

More noteworthy than von Harbou's rather florid prose is her recognition that, as much as the absence of light in the worker's underworld is at issue, so is the *excess* of light in the pleasure zone above. Unlike God's light which once served to clarify truth for Descartes in his moment of radical doubt, electric light not only illuminates but intoxicates, doubling and redoubling the city, recreating its streetscapes and squares as floating, dematerialised zones. In their "Technical Manifesto" of 1910, the Italian Futurist painters proclaimed:

Space no longer exists: the street pavement soaked by rain beneath the glare of electric lamps, becomes immensely deep and gapes to the very centre of the earth... (reprinted in Appollonio 1973: 28).

According to Fritz Lang, his film *Metropolis* was itself originally inspired by a visit to New York: "I first came to America briefly in 1924 and it made a great impression on me. The first evening, when we arrived, we were still enemy aliens so we couldn't leave the ship. It was docked somewhere on the West Side of New York. I looked into the streets—the glaring lights and the tall buildings—and there I conceived *Metropolis*".

Russian filmmaker Sergei Eisenstein's first impressions of New York, already famed as the brightest city in the world, register its vertiginous impact:

All sense of perspective and of realistic depth is washed away by a nocturnal sea of electric advertising. Far and near, small (in the *foreground*) and large (in the *background*), soaring aloft and dying away, racing and circling, bursting and vanishing—these lights tend to abolish all sense of real space, finally melting into a single plane of coloured light points and neon lines moving over a surface of black velvet sky. It was thus that people used to picture stars—as glittering nails hammered into the sky (Eisenstein 1963: 83).

Spectacular illumination of the urban landscape altered the accustomed mental image of the city, offering a new spatiality for modern experience. If by day poor sections of the city called out for reform, by night they could be redeemed by the power of light. Not only did lighting illuminate key urban landmarks, it effectively deleted others, casting unattractive areas into impenetrable darkness.

This capacity for architectural erasure was clearly appreciated by Andy Warhol, who shot his most notorious film, the eight- hour *Empire*, following the floodlighting of the Empire State Building in 1964. "The Empire State Building is a star", Warhol declared in his characteristic deadpan fashion, and for most of the film it literally is a star, continuously visible for more than seven hours in an unmoving frame. About 2.00 am the floodlights are switched off, and the last 45 minutes of the film are almost totally black. In an interview in 1975, Warhol commented:

> The best, most temporal way of making a building that I ever heard of is by making it with light. The Fascists did a lot of this "light architecture".
>
> If you build buildings with lights outside, you can make them indefinite, and then when you're through with using them you shut the lights off and they disappear (Angel 1994: 15).

Lights enable modern skyscrapers, skinned with glass curtain walls, to assume dazzling, indefinite forms, and then, finally, to disappear, as if their monumental forms are no more than a conjuror's trick. The immense possibilities for wholesale architectural substitution had been prefigured at the New York's World Fair of 1939, where the organisers decided that each building would have "a night appearance quite different from its daytime appearance", and so mandated that exterior lighting be built in to all pavilions as an integral part of their architectural design (Nye 1997: 125). Such a requirement dovetailed with General Electric's marketing of a new line of products called "luminous architectural elements" consisting of coves, grooves, recesses and coffers of the sort used in the Chrysler Building. The illumination of banners, sculptures, paintings and plaques at the expense of walls and roofs created dramatic contrasts, liberating details from their supporting context and structures from their physical surrounds. These examples enable us to appreciate the extent to which the generalisation of exterior lighting helped to engineer a new rhetoric of urban space. It provided the means through which the complexity of the modern city could be reduced to a few essential sites illuminated by floodlights, or grasped from above as a simplified pattern interspersed with unimportant blanks.

The ability to illuminate the cityscape in new ways introduced an important new dimension into urban design, one which belonged to neither architecture nor sculpture as traditionally understood. What emerges is a new urban environment increasingly characterised by the overlap of material and immaterial spatial regimes. With the proliferation of transparent and highly reflective surfaces, electric lighting creates a new perceptual matrix which overlays and even displaces the material spatiality of physical structures.

It is instructive to relate this new matrix of perception to the trajectories created by other contemporary technological developments. It has frequently been noted that the invention of the train literally changed the way that people saw the landscape (Schivelbusch 1986). Increased speed, combined with the elevation of the traveller and their immobilisation behind glass, altered the balance between foreground and distant elements, creating a convergence between voyager and voyeur. This tearing of the accustomed envelope of spatial continuity, resulting from the speed of the journey and the closeting of the traveller from physical interaction with the landscape, led Marcel Proust to compare train travel to metaphor, inasmuch as it "united two distant individualities of the world, took us from one name to another" (Kern 1983: 216).

If rapid vehicles heightened a sense of the journey as a staccato movement between what were increasingly conceptualised as discrete sites, this spatial awareness was re-enforced by the appearance of modern communication and media technologies. The live transmissions of telephony and radio interlinked physically separate spaces in novel ways, calling into question traditional spatial measures such as the solidity of walls and the security of distance. A 1912 editorial in the *New York Times* observed:

> All through the roar of the big city there are constantly speeding messages between people separated by vast distances and ... over housetops and even through the walls and buildings are words written by electricity. (Kern 1983: 64).

The electric city no longer provided a stable grid against which time and space could be measured in traditional terms. If electric light helped to turn the city into a promise capable of drawing the masses out of the countryside and across the oceans in their millions, the spatial organisation of the radiant city profoundly challenged customary understandings of place, boundary, dimension and locatedness. In doing so it crystallised one of the defining dilemmas of modernity: enhanced possibilities for individual freedom and self-invention are counterpointed by a growing sense of displacement and loss of orientation. The historic function of city design as a map of social and political order, and a repository of collective memory, started to give way to a new spatial organisation in which the co-ordinates of self, home and community would have to be plotted in new ways. In particular, any and every physical location now had to be reckoned in relation to its potential displacement by the activation of a circuit or the overlay of an image flow.

Long ago Gaston Bachelard reminded us that: "Everything which casts a light sees" (Schivelbusch 1988: 96). This situates the fundamental spatial ambivalence of electrification. With its unprecedented intensity, precision and control, it sets in motion a complex psycho-geography of seeing and being seen which is integral to the dynamic modern cityscape of promiscuous display and everyday voyeurism. Electric light not only illuminates the city, but changes the nature of urban spatial experience, creating oneiric "night cities" in which architecture seems to come alive. This uncanny dimension of the "technological sublime" prepared the ground for many contemporary developments in urbanism, including "smart buildings" in which spatial ambiance responds to the moods and movements of the inhabitants.

ENDNOTE

1. There are a number of exceptions, most notably the work of David Nye (1990). Also worth mentioning is Schivelbusch (1988), although its main focus is gas lighting.

REFERENCES

Angel, C. (ed.) (1994) *The Films of Andy Warhol: Part II* New York: Whitney Museum of American Art.

Appollonio, U. (ed.) (1973) *Futurist Manifestos* (transl. R. Brain et al), London: Thames and Hudson.

Bazerman, C. (1999) *The Languages of Edison's Light* Cambridge, Mass. and London: The MIT Press.

Benjamin, W. (1973) *Illuminations* (transl. H. Zohn) London: Fontana.

Le Corbusier (1964) [1935] *The Radiant City: Elements of a Doctrine of Urbanism to be used as the basis of our Machine-Age Civilization* (transl. P. Knight, E. Levieux & D Coltman) The Orion Press: New York.

Eisenstein, S. (1963) *The Film Sense* (transl. J. Leyda) London: Faber & Faber.

Freud, S. (1955) "The 'Uncanny'" (1919) in *The Standard Edition of the Complete Psychological Works of Sigmund Freud* (transl. under the general editorship of J. Strachey 1955) XVII London: The Hogarth Press and the Institute of Psychoanalysis 219–52.

von Harbou, T. (no date) *Metropolis* London: The Reader's Library.

Kern, S. (1983) *The Culture of Time and Space 1880–1918* Cambridge, Mass.: Harvard University Press.

Koolhass, R. (1994) *Delirious New York: A Retroactive Manifesto for Manhattan*,Rotterdam: 010 Publishers.

McQuire, S. (1997) "The Uncanny Home" in *Paradoxa: Studies in World Literary Genres*, 3/. 3–4 527–38.

------------. (1998) *Visions of Modernity: Representation, Memory, Time and Space in the Age of the Camera* London: Sage.

Marx, L. (1965) *The Machine in the Garden: Technology and the Pastoral Ideal in America* Oxford University Press.

Morus, I. (1998) *Frankenstein's Children: Electricity, Exhibition and Experiment in Early Nineteenth Century London* Princeton, NJ, Princeton: University Press.

Nye, D. (1990) *Electrifying America: Social Meanings of New Technology 1880–1940* Cambridge, Mass. and London: MIT Press.

Nye, D. (1997) *Narratives and Spaces: Technology and the Construction of American Culture* Exeter: University of Exeter Press.

Robinson, C. and Bletter, R. (1975) *Skyscraper Style: Art Deco New York* Oxford University Press.

Schivelbusch, W. (1986) *The Railway Journey: The Industrialization of Time and Space in the Nineteenth Century* Berkeley: University of California Press.

Schivelbusch, W. (1988) *Disenchanted Night: The Industrialization of Light in the Nineteenth Century* (transl. A. Davies) Oxford: Berg.

Woroszylski, W. (1971) *The Life of Mayakovski* (transl. B. Taborski) New York: Orion.

[RACHEL MOORE]

Love machines

This chapter is part of a project entitled, "In the Film Archive of Natural History", the aim of which is to work out why various outmoded technologies, degraded images, and old films hold particular sway today. This problem has been suggested to me by two very different things. One is the oldness of new films—the large degree to which recent avant-garde cinema deploys old footage, old technologies, old objects, old settings, as the recent work of Martin Arnold, Janie Geiser, Ross Harley, Ken Jacobs, Lewis Klahr, and the Quay Brothers, for example, all re-work the old into the new in important ways. The other is the surprising newness I experienced while watching archival footage.

Old films and footage are fascinating not just because they are old but because they look old. With the enormous span of time now embedded in the very grain of the celluloid, they retouch, in a sensate way, the strange and familiar longing for the archaic past which lies at the heart of the modern dilemma. Walter Benjamin's suggestion (1983–4: 6)—that when delving into the secrets of modernity, its technology for instance, the archaic is never that far off—grows palpable when watching film from the archives.

To begin to get a purchase on how old films, in particular, touch us, it is instructive to look at how technology functions within films. The power of degraded technology to create intimacy does not go unnoticed by filmmakers today where its use extends from the avant-garde to popular cinema. One instance from the avant-garde is Sadie Benning's pixelvision, which uses a low-tech video image such that the pixels appear to form a visible permeable skin that dissolves the border between the viewer and the film screen from within. From popular cinema, in the film *Gods and Monsters* (Condon 1998), the two central characters watch the blurry images of the movie *Bride of Frankenstein* (Whale 1935), on separate television sets, one in a bar, the other in his living room. Come morning when the two meet, their friendship has been uncannily cemented. In the film *Central Station* (Selles 1998), the woman sitting in a bus as she is leaving the child, and the child who is left behind, each hold up tiny plastic slide viewers simultaneously, and look at a single slide picture taken at a high point along their journey, thus bringing their fraught relationship to a joyful conclusion. These scenes use the charm of outmoded technology to secure an intimacy between characters otherwise out of physical reach. The intimacy effect of degraded technology, the bad televisions and the simple slide viewers, operates within films themselves, but the old film we watch also draws out a similar enervating effect when the material detritus of time, in the form of patterns, changes of colour, and aging's various haphazard scars surface to suggest just one reason. To further understand the nature of the intimacy technology evokes, this chapter focuses on one way technology provokes intimacy: how people fall in love in the movies (on the film screen that is).

As we watch old films in which people are shown getting to know each other and falling in love, the magic moment in which love is mutually acknowledged is invariably conferred by some kind of attention to technology. *Storm in a Teacup* (Saville, Dalrymple 1937), perhaps one of England's first animal rights pictures, is set in a Scottish village. It is the story of a father Provost Gow (Cecil Parker) and his daughter Victoria (Vivien Leigh), a newspaper reporter from England named Frank (Rex Harrison), the beleaguered Irish ice-cream vendor Mrs Hagerty (Sara Allgood), and her dog "Patsy" ("Scruffy"). Frank and Victoria's concerns overlap finally over the fate of the dog that her father has ordered to be put to death. Victoria is exasperated by Frank who has not only failed to charm her, but has insulted her father to whom she is devoted. Leaving the father's house after a row with the father, Victoria follows Frank into an amusement arcade, just the sort of fairground of attractions where early cinema was once found. As they wander through, they are distracted by different amusements, but nonetheless tend to a conversation as they put coins in slots and play with the various games on display. While Victoria's back is turned, Frank stamps Victoria's name in a metal press; they continue to talk while Frank vents his anger swinging the pile driver. She asks him a question and Frank responds triumphantly, "Frank! you called me Frank, that's the first time you've ever spoken my name!" From out of nowhere, a life-sized robot in humanoid form whose eyes spontaneously light up, blurts out Frank's weight. They turn from the glowing robot to gaze upon each other's now likewise radiant faces. Continuing their affectionate debate, Victoria accuses Frank of being stubborn while Frank looks into a mutoscope, the only attraction in this episode to whose name we are not privy.[1] She says, "stubbornness is as old as the pigs," to which he responds gesturing to the viewer, "so is this". Victoria takes his arm, the encounter ends with his handing her a strip of metal with the words "I love you, Victoria" pressed into it; thus the rest of the film can be devoted to the fate of the dog.

I offer this example not because it is rare, but because it is so extreme. The robot—by turns primitive *and* futuristic—would appear outlandish in any temporal context. In technological guise, this cupid leaves the couple awestruck and vulnerable to a likewise primal passion, one as old as the pigs. Technology's role as Ariel does not stop there. Victoria's father, who has been running a political campaign throughout the film, is finally publicly shamed for his insistence that the unlicensed dog be killed, so he rights the wrong he has done to ensure his successful candidacy. A pompous man nonetheless, his campaign speech is ennobled by a mural of Robert The Bruce depicted here in a kilt made of feathers, bare-chested in his warrior cave watching the spider from which he drew rebounding strength. The Bruce dominates the frame from behind, while a broadcasting microphone in the foreground of the film frame and thus very large, covers the crotch area of the portrait's kilt with noteworthy precision. The father struts mightily in his ritual garb, his kilt adorned by a large feathered sporran with horns curling forth graphically matching The Bruce's feathered kilt. His voice carries to the next frame issuing from a car radio adorned by a mere glimpse of Victoria's leg as she and Frank make their honeymoon getaway. Modern technology stirs up love between the character's very legs twice: bestowing the father's blessing via the masterfully placed microphone on the pairing of Victoria and Frank where the radio speaker emanates its heat on the eager limbs which wait just below the gear stick.

Though hardly a great film, I use *Storm in a Teacup* as an illustration of technology as a primitive, magical conduit for love because it is so obvious. From the robot's scene in the arcade and from The Bruce as he derives an object lesson from the spider, the archaic travels in a technological flash to the now.

A better-known and far subtler example would be Leo McCarey's *Love Affair* (1939) in which, after much dipping and dodging aboard an ocean liner, Irene Dunne and Charles Boyer stand on the deck and agree that their romance is serious. As they move into New York harbour, the reigning American symbol of technological industry, the Empire State Building stands tall behind them in the centre of the frame.[2] Later, the building serves a narrative function, but here, in the moment when they are assuredly in love and they turn to look at the towering structure, its grandeur returns their gaze as inspiration. And love needs, at least in represented form, inspiration. To understand how modern technology got the role of Cupid, I will briefly consider two well-worn apophthegms: that love is magic, and that technology is magic.

With "the voice of the turtle" in the *Song of Solomon*, nature confers the moment of love's ascendance: "The flowers appear on the earth; the time of the singing of birds is come, and the voice of the turtle is heard in our land" (2: 12, King James Version). What sounds here like nature's blessing also marks a moment when nature colludes with Eros as when the appearance of the snake, for example, startles Eve's composure in Preston Sturges' *The Lady Eve* (Paramount 1941), effectively tripping her into love. In pre-cinematic, pre-modern technological forms, like Shakespeare's *The Tempest*, among Ariel's magical manoeuvres on Prospero's account is the charming of Miranda and Ferdinand such that they are immediately devoted to one another. Cupid's shaft made the magic juice Oberon uses to make the various couplings in *A Midsummer Night's Dream* and Hermione's statue is animated by Leontes' adoration in *A Winter's Tale*. The story of Tristan and Iseult's love potion and the horns that finally blast down the Walls of Jericho in *It Happened One Night* (Frank Capra 1934) alike accede to the notion that romantic love requires a magical trigger that comes from outside the lovers' bodies, and is powerful enough to affect the bodies' every sense.

In much the same way that magic became the link between the word and its meaning for such thinkers as Walter Benjamin, Eros for Plato was endowed with the status of the link between existence and the essence of beings (Couliano 1987: 4). Eros, the "pneuma" that transmits between beings, is magic. Couliano, in his Renaissance study of *Eros and Magic* cites Ficino's *Amore*: "The whole power of Magic is founded on Eros. The way Magic works is to bring things together through their inherent similarity" (1987: 87). He goes on to describe love as a kind of sympathetic magic akin to the way the "organs interact, favor each other". Ficino concludes, "From this relationship is born Eros, which is common to them all; from this Eros is born their mutual rapprochement, wherein resides true Magic" (1987: 87). If magic is love, Couliano argues, love is magic. As the animus that communicates between the two bodies, magic merges one person with another.

Irving Singer connects the idea that love is magic through the idealist and what he calls the realist traditions to insist that still today love is something over and above, an excess that maintains its magical status perhaps from its very gratuitousness. With the concept of bestowal, however, he insists that it comes not from outside—that is merely the stimulus—but generates from within a person. "Love supplements the human search for value with a capacity for bestowing it gratuitously." For Singer, in loving one "subordinate[s] his purposive attitudes" and "transforms himself into a being who enjoys the act of bestowing". Love, for Singer, is by nature excessive, in the sense that it exceeds utility. Singer concludes: "There is something magical about this [bestowal], as we know from legends in which the transformations of love are effected by a philter or a wand. In making another person valuable by developing a certain disposition within oneself, the lover performs in the world of feeling something comparable to what the alchemist does in the world of matter" (Singer 1991: 218). While Singer argues that the magical transformation occurs within the person, stories, folklore, plays, and indeed his very rhetoric rely on some form of magical Cupid to illustrate love. The idea that love requires a trigger from outside gives force to Spinoza's definition of love: "love is an elation accompanied by the idea of an external cause" (Curley 1994: 189). Magic is the idea, at least, of love's external cause. In modern times, magic inhabits technology to produce the necessary elation of the body's senses.

Aristophanes' contribution to the *Symposium* describes the birth of love itself as a magic act in which Zeus had Apollo bisect the original human being that was a round, two-personed figure, turned their heads around, stretched the severed skin together over what became their front and pulled it taut with a drawstring. With this act, Zeus set us off looking for our lost half, and thus begins love as we know it today, love founded on lack, on a lost wholeness. It is also the birth of physical eroticism, as "previously begetting and birth had been accomplished by emission on to the ground, as is the case with grasshoppers" (Plato 1951: 61). "It is from this distant epoch", Plato reports Aristophanes to have said, "then that we may date the innate love which human beings feel for one another, the love which restores us to our ancient state by attempting to weld two beings into one and to heal the wounds which humanity suffered" (*Symposium* 62).

In Ann Carson's *Eros the Bittersweet*, breaks, edges, and taut triangulations are the basis for her reading of lyric poetry in which eros is the impossible possible connection we understand as the reach of desire: "For where eros is lack", she writes, "its activation calls for three structural components—lover, beloved and that which comes between them. They are three points of transformation on a circuit of possible relationship, electrified by desires so that they touch, not touching" (1998: 16). Love is an impossible thing. It requires the heavens to shake, the snake to jolt Eve into a state of terror, Cupid's poisoned arrow to electrify desire, so as to perform the impossible touch which does not touch.

The archaic lyric poets were not only grappling with the triangulation of desire, they were also, according to Carson, contending for the first time with writing. Unlike the spoken word, writing, and especially Greek writing, has edges, breaks, clearly defined borders. She suggests that "If the presence or absence of literacy affects the way a person regards his own body, sense and self, that effect will significantly influence erotic life. It is in the poetry of those who were first exposed to a written alphabet and the demands of literacy that we encounter deliberate meditation upon the self, especially in the context of erotic desire. The singular intensity with which these poets insist on conceiving eros as lack may reflect, in some degree, that exposure" 1998: 44). The edges of words were part and parcel of the redefinition of a boundless social self, reined in from social speech's amorphous expansiveness to the confines of writing. To such an individual who "appreciates that he alone is responsible for the content and coherence of his person, an influx like eros becomes a concrete personal threat" (1998: 45). "So in the lyric poets", Carson surmises, "love is something that assaults or invades the body of the lover to wrest control of it from him, a personal struggle of will and physique between the god and his victim" (1998: 45). The written word operates in the same realm of absent presence as eros. Like eros, writing "is an act in which the mind reaches out from what is present and actual to something else" (1998: 61).

So writing, for the archaic lyric poets at least, is the technology that performs the impossible labours of love. Finally, to love partakes of the same reach of imagination as to know, and it is no coincidence I think that the customary way to represent someone getting an idea, is to make a light bulb appear. Carson writes, "I would like to grasp why it is that these two activities, falling in love and coming to know make me feel genuinely alive. There is something like an electrification in them" (1998: 70). In archaic times then, Zeus' magic and the technology of writing create the fissures to which desire continues to attend to this day.

The modern figure for trickery and ingeniousness, revelation and concealment, is technology. Technology reveals its cunning and trickery most clearly when cast against the un-modern: those savages who are amusingly mystified by bottle openers, cream separators, telescopes, binoculars, telephones, theodolites, and, of course, the record-player.[3] The last of these has endured from Flaherty's *Nanook of the North* of 1921 to the 1999 version of *A Midsummer Night's Dream* (Hoffman) for a brief show of primitive wonder at the machine when the Inuit Nanook and the fairy examine the record and then pick up the record as if to bite it. These examples illustrate the projection of a magical status of technology onto the primitive, which actually more accurately reflects the modern person's attitude. Thrown into primitive relief, technology really does not do anything but instead shares the trickiness and wiliness which are otherwise leached out of us through technology.

Marshall Sahlins once said that the more primitive the technology, the more sophisticated the operator. If this observation reflects something of the modern attitude towards technology (and it is indeed borne out by past and present filmic interest in primitive hunting, craftwork, and the like), our technology today makes us feel a little less than clever. Thus the fear in Heidegger's observation that "the will to mastery becomes all the more urgent the more technology threatens to slip from human control" (1993: 313). Technology, which we are always trying to get a grip on, is as elusive as it is ubiquitous.

More importantly for Heidegger, technology is "no mere means" but rather it is as he puts it, "a way of revealing". "It reveals", he says, "whatever does not bring itself forth and does not yet lie here before us, whatever can look and turn out now one way, and now another." That is to say it refers not to the manufacture of something but rather what he calls the "bringing forth" of something (say, a ship or a house) which is already a finished thing envisioned as completed "but is revealed through technology" (1993: 319–20). This same process occurs as physis, "The arising of something out of itself" when a blossom, for example, bursts into bloom (1993: 317). And here it is difficult not to recall the cinematic mileage that has been achieved from stop-action photographs and close-up films of nature's marvels. For it is in the process of revealing that is technology itself where the camera is in its most cinematic element. So Sappho's nature—"mountain winds descending on oaks"—is not that far off from the robot at the fairground show. The technologies that feature in films are no mere means—that is to say their primary importance is not to function as machines, rather, they stand by to *reveal* love.

Since technology primarily stands as a way of revealing, I have discounted the telephone, the internet, and the like, when they function as a means of communication. A scene from *They Drive by Night* (Raoul Walsh 1940, US Warner Brothers), illustrates the clarity with which this distinction obtains, however intuitively. Joe (George Raft), in San Francisco decides that Cassie (Anne Sheridan), in Los Angeles, is the one for him so he calls her. On the left side of the frame Joe nuzzles up to the public telephone on the wall, and Cassie, on the right, faces left to coo into her similar telephone and complete the symmetry. Suddenly, from the middle of the frame, a superimposed image of a telegraph pole with four horizontal scaffolds holding wires and insulators erupts, Vertov-style. Moreover, it hums loudly. The telephones are merely functional objects, but the towering telegraph pole, like a spinal column and rib cage zinging a message through the nervous system, is technology in full form, interrupting the scene to stand mightily and a little eerily here as Cupid. The eeriness which features in this example derives from another quality of technology which its representation as both an apparatus and a skeleton suggests: the double sense Heidegger intends with the word *Gestell* which he uses to name and thus distinguish the essence of *modern* technology. For, in addition to the concealing–revealing which characterises all technology, modern technology "sets upon nature and challenges it to yield" (1993: 326).

This sense, recently taken up by environmentalists, understands technology's uniquely modern capacity to sap and to store energy as one that effectively turns nature into a servant of technology. Here technology, in the form of the telegraph pole, stands between love and marriage. As such, it stands as the energy that takes the natural, wild and erotic love, and turns it into love's cultural form, marriage.[4] The superimposed telegraph pole is both the armature for and the skeletal remains of erotic love. Joe and Cassie, the lovers on the brink of marriage throughout the story, are wedged between two unfortunate models for romance. In this film marriage is either deadly (if your partner is Ida Lupino), or dowdy (as with the dull and depressing marriage of Gale Page to Humphrey Bogart). For all the flourish of Walsh's kinesthetic gesture, its wizardry in the end merely reifies rather than enchants. Like a hydroelectric plant that harnesses a river's power and transforms it into metered current, technology here churns nature out as second nature. This too is a magic act. Technology's premier feature is the secreting of its transformative power which makes it appear now frightening, now wondrous, but always as strange. In this, technology shares with the commodity fetish the unnatural role of a magical attraction.

In Max Ophul's *Letter from an Unknown Woman* (1948), the lovers visit the Viennese fairground the Prater and take a simulated carriage-ride trip around the world to seal their romantic fate. (And much could be made of the role of immobile carriages and cars as vehicles for love; for, from Emma Bovary's trysts to the teenager's flat tyre, it is only when something stops working that it reveals itself fully.) Technology here is an attraction like the amusement park ride, not a means of manufacture. Similarly, the "cinema of attractions" (Gunning 1986: 63–70) as opposed to the absorptive pull of narrative-driven cinema rejects its functional element in favour of cinema's magical quality. Customarily, when films simulate early viewers' responses to "new" technology, they highlight the magical qualities of cinema by flaunting its technological prowess.

Francis Ford Coppola's *Dracula* (1992) deployed the titillation and fear often projected onto first contact with the cinema and other related technology in the service of seduction. Count Dracula takes Wilhelmina through an arcade of attractions in turn-of-the-century London streets to view the cinematograph because, he says, "I understand it is a wonder of the civilized world." On the screen two scantily clad women are seated on a man's lap. They, much to his chagrin, disappear and are replaced by a fully clothed woman. "Astounding", says Dracula, "There are no limits to science." Mina starts to leave. A film of a train moving towards them in negative is on the screen in the back of the frame. "Do not fear me", Dracula says desperately and drags her to a cloistered seat in the rear. He holds her, caresses her, while on the screen a naked woman faces the screen, turns around and walks away. Fascinated and afraid—Mina is utterly seduced. She runs through the back stage of an elaborate shadow play, only to be frightened by his white dog (which was black a few scenes earlier) and thus begins their final seduction scene in which both lasciviously caress the unnatural dog. Carson's description of Eros' effect could not be more precise: "Consistently throughout the Greek lyric corpus, as well as in the poetry of tragedy and comedy, eros is an experience that assaults the lover from without and proceeds to take control of his body, his mind and the quality of his life. Eros comes out of nowhere, on wings, to invest the lover, to deprive his body of vital organs and material substance, to enfeeble his mind and distort its thinking, to replace normal conditions of health and sanity with disease and madness" (1998: 148).

The film that forms the backdrop for the scene's climax and the birth of their romantic bondage is again that image of the train advancing from the darkness. Haunting the entire sequence, this film is the 1901 Biograph *The Ghost Train* (or more precisely, its replica) in which the train The Empire Express is shown in negative form which makes for the frightful image of a white train zooming out of a black world. With its placement here in the smoky arcade, it markedly echoes cinema's primal scene, the showing of Lumieres' *Arrival of a Train* in the Grand Café (France 1895) which was then greeted with such "astonishment" (Gunning 1989: 3). The astonishment accorded to the reception of "primitive cinema", of the "cinema of attractions", unleashes a similar wonder and vulnerability here. The clips of film Coppola chooses with which to impart Eros' presence not only foreground cinema as a novel attraction, they are also examples of the specific things—like jump cuts, negative exposure—that only film can do. The technological aspect of cinema in its magical sense, even more than cuddling up in the back row of a cinema, is what, for Dracula at least, makes romance at a film show.

Such purely cinematic moments, such instances of cinematic physis are precisely those that sustained both Benjamin's and Kracauer's interest in films. The materiality of the image addresses the viewer as "corporeal-material being" seizing it, as Kracauer put it "with skin and hair".[5] As with Benjamin's "innervation", the image grabs the senses, and pulls them into direct contact: "The material elements that present themselves in film", Miriam Hansen puts it, "directly stimulate the material layers of the human being: his nerves, his senses, his entire physiological substance" (1993: 452). The material elements of cinema: slow, fast motion, jump cuts, angles of all directions and frames in all sizes, coloured filters, and the very aging of the film itself—all serve to give technology a key part in a movie, any movie.

In the film *Get Shorty*, Chilly Palmer (John Travolta) chooses a film by a director known for utilising and expanding cinema's formal properties, Orson Welles, to combat Karen Flores' (Rene Russo) romantic scepticism by inviting her to meet him there. As in *Dracula* the love scene at the cinema is inaugurated by a shot of the film being shown rather than the spectators watching it, reliving, I suspect, in both instances, a fleeting moment of "first contact" with the moving image. Welles' signature low-angle shot shows Charlton Heston with a gun disproportionately enlarged due to foreshortening pointed at him saying, "How could you arrest me here, this is my country." A high-angle close-up of Orson Welles responds for the showdown that transpires. With the sound from the film still audible we finally see Chilly, so engrossed in the scene that he is literally on the edge of his seat, when Karen ambles into the cinema.

She spots him but takes a seat far away so she can watch both Chilly and the end of *Touch of Evil* (1958). Anticipating the dialogue, Chilly says, "that's the second bullet I stopped for you", echoed immediately on the soundtrack. Karen smiles in much the same way as one does watching a child with an animal at the zoo. A serious film enthusiast, Chilly knows the film by heart and continues his recitation until he cannot wait anymore, "you're going down, Orson", while Welles' enormous body hits the water. Chilly beams at Marlene Dietrich's (Tana in the film) gorgeous close-up lingering on the screen and this time the viewer's identification shot is matched to Karen at full attention to Tana's vacant eulogy: "he was some kind of man". Tana's "Adios" is the last word, the pianola's brittle trills escort her lone walk along the road where monstrous oil rigs foreground deep and dark space. Chilly's eyes, utterly besotted, look left to right as if to say, "I don't believe it!" Clearly the opposite of his unflappable personae; he even elbows a stranger to extol the film's greatness. The pianola's refrain from *Touch of Evil*'s end[6] accompanies the two out onto the street as they talk and Karen ends the conversation saying, "this may just work out".

The pianola music that envelops the film with the film within the film comes from Tana's place where it played when Quinlan called on her in *Touch of Evil*. Their conversation there began with Quinlan's remark, "That music brings back memories." Putting a cigar dead-centre in her mouth, Tana replied, "Pianola music, the customers go for it. It's so old it's new. We got the television too, we run movies." Just as Tana brutally distinguishes the market charm of the pianola from the nostalgia Quinlan would prefer, old films are not vague mnemonics of times past now viewed through the scrim of experience. Rather, they are so old, they are new. As if unwittingly awakening to cinema's first performance, a film's oldness instantly invokes its novelty and newness. Old films, with their different styles which now stand out to us as artifice and their more severe surface scars, flaunt their materiality. The newness of the old "retouches the real with the real" (Bresson 1986: 44) "as dust on a diamond reveals its transparency" as Andre Bazin said in another context. (1961: 131)

Touch of Evil is the raunchiest film Welles, or perhaps anyone else, ever made and hardly a story to elicit the broad smile on Chilly's face at its end. Like Flaubert's idea for *Salambo* he notes in his letters it is a film that is pure style. Indeed there is a similar sinister delight at the stylistic punishment such myopic subjects as Emma Bovary and Heston's Vargas undergo. This is not the amusement-filled gallery of early cinema's crude attractions to which Dracula subjected Mina. The use of this film ups the technological ante. This is the cinema of tape recorders used to supply diegetic sound, the longest take in film history, violently ironic cross-cutting, angles so acute a gun dwarfs Charlton Heston's face. Chilly is taken in by the audacity of pure style. And Karen is seduced by the way the film works Chilly over such that, like love itself, the old becomes new. This subtle and intricate use of technology as stylistic prowess to touch off romance is a far cry from clichés of waves breaking, suns setting, or breezes blowing. Those could only be parodic for two people who, as defined by the film, know a scam when they see one. If anything differentiates the technological Eros from natural Eros it is a degree of guile.

It could be added that the more sophisticated the couple, the more wily their technological romantic muse. The technological magical cupid appears not at first encounter, but after a series of interchanges in which Spinoza might say the characters develop adequate ideas. Like Karen, the most promising lovers spend a fair amount of time observing each other. As Amelie Rorty puts it, "passive elation moves to active elation and she develops adequate ideas, as she understands what within her he loves and what in him she loves" (1991: 352–71).

In Stanley Cavell's study, *Pursuits of Happiness* (1981), he creates a genre he calls *The Hollywood Comedy of Remarriage*. In much the same way Cavell understands "remarriage" the instances in which the intimacy–effect of technology surfaces with utmost subtlety are mature romances. In Leo McCarey's *The Awful Truth* (1937) Jerry (Cary Grant) and Lucy (Irene Dunne) go through the peregrinations of separation and divorce only to find themselves in separate bedrooms with a communicating door at the film's end. A black cat bars the door which the couple both wish will open as a cuckoo clock on the wall ticks toward the hour of midnight when their divorce will become final. When the clock strikes quarter past eleven, "Two childlike figurines", as Stanley Cavell describes them, "somewhere between live figures and automatons, appear from adjacent doorways and skip back in, the two doors closing during the last chimes" (1981: 256). These figurines appear twice at quarter-hour intervals, skipping out, turning around, and skipping back into the clock to mark the time.

When the door between Jerry and Lucy's room opens for the last time they reconcile. In the one room with a closed door, Lucy smiles at Jerry from her comfortable bed. Thanks, only in part, to the censorship of the times, this is the last we see of Jerry and Lucy. We are spared their final embrace, the literal truth, what Nietzsche has told us is merely dead or fossilised metaphor. The finale is performed by the cuckoo clock, striking the last interval before the midnight hour and the two figurines reappear from their separate doors for the third time, skipping out, turning round and starting back into their separate doors. But this time the boy figurine jumps his track and joins the other to exit through one door, accompanying the girl back into the clock. The figurines, neither human nor not human, are technological tricksters of a high order; wily enough to impart love's magic upon a couple who are, after all, already married. Like ornamental gnomes of gardens, these figures rest uneasily in the present, anxious to scurry back into mythic history with or without imparting their charms.

Within the film frame, person and thing alike are levelled to the status of an image, and it is in the image where metaphor and the mythic find a purchase. The cinematic object purloins subjectivity and hands it over to the world of things. The earrings, for an extreme example, in Ophuls' *The earrings of Madame de…*(1953) accrue more value with each exchange, while draining the vitality of the lovers to the point of death, such that the once forgettable objects, the earrings, become brilliant fetishes so powerful they must be gifted finally to God. These images, with their thing-like status, in turn function not merely as ciphers but as glistening fetishes, fetishes reanimated by the very subjectivity lost to the object. What distinguishes these technologically charged objects from mere commodity fetishism is the way they can bring back the mythic—in this case, the primordial knowledge of love.

"Only a thoughtless observer", wrote Benjamin, "would deny that there are correspondences between the world of modern technology and the archaic symbolic world of mythology." (1983–4: 6). The mythic takes over in the realm of imagination to attend what cannot otherwise be known; the mythic, wrote Kierkegaard, "allows one to imagine and to make good the loss" (1989: 107). And the mythic always appears as image. Our preeminent form of second nature, the cinematic image, becomes, a dialectical image. Cinema holds a storehouse of these images, dormant, ready to burst forth as dialectical images.[7]

Cinema itself rests uneasily in the present. This is true no less for the humble spectator as the film rushes on in front of your eyes as it is for the film preservationist. From the minute they are produced, films change, deteriorate. "Let's face it", writes Paolo Usai, "the most stable medium known by human civilization is ceramic. Glass is all right. Stone may be affected by pollution. Canvas and wood have some problems. Something can be done about paper and frescoes, but the gelatin emulsion of a film has been for a hundred years a thin layer of organic material. Gelatin. Animal bones, crushed and melted into a semitransparent layer interspersed with crystals of silver salts. It won't last. It can't" (Usai 1999: 43–4). Like a living creature, celluloid smells putrid as it decays. In the end it is reduced to a powder.

The film archive of natural history is truly "a charnel house of rotting interiorities", to cite Lukacs' description of second nature (1971a: 64). These reified strips of time may remain just that, nonetheless cinema remains an archive from which we can retrieve and reinvigorate images to suit our more mythic purposes. At once new and old, these images are as unstable as they are potentially magical and above all indifferent to the present. The magic cuckoo clock ticks away the time in *The Awful Truth* "much like an antediluvian fossil reminding us of another kind of life that doubt has eroded" (Kierkegaard 1989: 103). The garden gnomes—skipping in, and skipping out—their indifference and their presence, taken together perhaps accounts for Cavell's parting remark to these figurines when he says, "this clock is a mythological object", an object, what is more, "only available in the cinema".

It is the sense of the mythic, standing by ready to break forth, but just as prepared to leave, which makes watching old films a real experience. The film archive of natural history is that storehouse of images which allows us to imagine, and to make good the loss.

ENDNOTES

1. The use of the mutoscope here is particularly coy, for this is where the more explicit pornography was often shown. So the director is revealing a technology which is revealing the sex that is concealed in the film.
2. Of course, the Empire State Building can be variously interpreted. To William Burroughs, for example, it always looked like a hypodermic needle.
3. Cream separator, *The Old and the New* (Eisenstein 1929); telescope, *Dances with Wolves* (Costner 1990); binocular, Reynolds 1991); theodelite, *Dersu Usala* (this is a complex example, for it is the surveyor who marvels, to the point of making drawings of it, at the way *Dersu Usala* (Kurosawa 1975) uses the theodelite as the structure to build a hut); telephone *Mildred Pierce* (Curtiz 1945).
4. Georg Lukacs (1971: 100): "There is no natural form in which human relations can be cast... without their being subjected increasingly to this reifying process. We need only think of marriage, and without troubling to point to the developments of the nineteenth century we can remind ourselves of the way in which Kant...described the situation: 'Sexual community', he says, 'is the reciprocal use made by one person of the sexual organs and faculties of another...marriage...it's the union of two people of different sexes with a view to the mutual possessions of each other's sexual attributes for the duration of their lives.'"
5. Miriam Hansen (1993: 4) paraphrasing Kracauer in "With Skin and Hair".
6. The pianola playing here is, unlike most of the sound in *Touch of Evil*, non-diegetic, and thus free to roam outside the cinema.
7. My account of the cinematic image's status as commodity fetish is in *Savage Theory* Duke 2000 Chapters 5, 6, and 7.

REFERENCES

Bazin, Andre (1967) *What is Cinema?* (transl. Hugh Gray) Berkeley: University of California Press.

Benjamin, Walter (1983–4) "N [Theoretics of Knowledge; Theory of Progress]" *The Philosophical Forum* Xv/ 1–2 (Fall–Winter).

The Bible King James Version *Song of Solomon* 2–12.

Bresson, Robert (1986) *Notes on the Cinematographer* (transl. Jonathan Griffin) London: Quartet Books.

Carson, Ann (1998) *Eros the Bittersweet* Normal, IL: Dalkey Archive Press.

Cavell, Stanley (1981) *Pursuits of Happiness* Harvard Film Studies.

Couliano, Joan P. (1987) *Eros and Magic in the Renaissance* (transl. Margaret Cook) Chicago: University of Chicago Press.

Curley, Edwin (ed. and transl.) (1994) *A Spinoza Reader* Princeton: Princeton University Press.

Gunning, Tom (1986) "Cinema of Attraction: Early Film, Its Spectator and the Avant-Garde" *Wide Angle* 8/3–4.

---------------- (1989) "An Aesthetic of Astonishment: Early Film and the (In)credulous Spectator" *Art & Text* 34 (Spring) 31.

Hansen, Miriam (1993) "With Skin and Hair" *Critical Inquiry* no. 19, Spring.

Heidegger, Martin (1993) "The Question Concerning Technology" in *Selected Writings* (ed. David Farrell Krell) San Francisco: Harper.

Kierkegaard, Soren (1989) *The Concept of Irony* (transl. Howard V. and Edna H. Hong) Princeton: Princeton University Press.

Lukacs, Georg (1971a) *History and Class Consciousness* (transl. Rodney Livingstone) Cambridge, Mass.: MIT Press.

Lukacs, Georg (1971b) *Theory of the Novel* (transl. Anna Bostock) Cambridge, Mass.: MIT Press.

Plato (1951) *The Symposium* (transl. Walter Hamilton) London: Penguin Books.

Rorty, Amelie (1991) "Spinoza on the Pathos of Idolatrous Love and the Hilarity of True Love" in *The Philosophy of Erotic Love* Lawrence: University of Kansas Press.

Singer, Irving (1991) "The Nature of Love" in *The Philosophy of Erotic Love* Lawrence: University of Kansas Press.

Usai, Paolo Cherchi (1999) "A Model Image, iv. Decay Cinema: The art and aesthetics of moving image destruction" in *Stanford Humanities Review* 7/2. 1–49, 43–4.

[ANNETTE HAMILTON]

The uncanny in object relations, or, love with the machine

This paper moves in a zone at the edges of several disciplines: somewhere between anthropology, philosophy and psychoanalysis, and in other places as well: in the lonely frightening spaces of early infant experiences, across the transitional boundaries between wilderness and civilisation, in our present only-just inhabited world of new technologies and primitive dreams of the Will-to-Power.

How are we to understand our relationships with the technologised lifespaces surrounding us, with the programs which require new attributes of the wetware, restructured cognitive skills and new forms of subjectivity appropriate to the machine-world? The uncanny qualities of photography and cinema were often noted in the nineteenth century, a kind of ghostliness in two dimensions. When early photographers tried to capture images of newly encountered tribal peoples, the people fled. Although it has become a cliché of the photographer and the native, it is nonetheless true that many tribal people did fear the effect of the machine as it faced them. When they were shown the results in a photograph, they could not at first decipher the two-dimensional black and white images. The camera effected a kind of capture: the photographer "shot" the people, and, it was thought, stole their spirits or "souls". Almost without exception, pre-modern humans believed something animated them, an essential element which arrived at conception or birth or shortly after and departed with death. In the absence of the spirit, the body was "dead meat". Ghosts and spirits: animating substances in dead flesh, immaterial elements of once-existent Beings. Yet today we are comfortable with every conceivable form of representation, with seeing things which cannot or should not exist such as dinosaurs and aliens, or experiencing with our own senses (at the cinema or with the computer) demons and cyborgs and every conceivable, imaginable creature and thoroughly enjoying it without the slightest concern for our own supposed inner essences. How have we become so comfortable, so at-ease, with technological "things", so that far from fearing or avoiding them, we are willing to do whatever is necessary to obtain them—even hand-held telephones which can send photographs to our refrigerators?

While these relationships can sometimes cause a kind of vertigo, a slipping into a condition where the desire to resist is as strong as the desire to yield, I want to suggest that this in fact is typical of the experience of relationship between humans and their Things. New technologies always appear alarming and dangerous: in Australia, finely worked stone knives with a blade of trigonal form manufactured from quartzite were traded into the Arrernte lands and from there south into the Western Desert. Whereas the knives were utilitarian instruments in their place of origin, by the time they reached the Western Desert they were wrapped and kept apart from any everyday activity, to be revealed only on the most sacred-secret occasions—they had become sacred objects (Spencer and Gillen 1968: 593; Hamilton 1980: 114–15). The movement between sacredness and danger, and everyday appropriation, seems likely to have been a condition of the human relation with technology throughout our history.

The presence of Things, of Objects, has oscillated in our intellectual and social practices. In the nineteenth century, "primitive" and exotic objects (African masks, Japanese screens, Chinese porcelains) rushed into prominence as objects of desire, and the study of "material culture" emerged, an aesthetic anthropology which focused entirely on the Thing, its appearance, treatment and method of manufacture. Museums were filled up with Things, ripped out of any context and stripped of any original meaning, objects for Westerners to collect and contemplate. In the mid-twentieth century Things disappeared, partly into the political economy perspective (Things were the product of alienated labour, and the like) and partly into science and technology studies. They reappeared in the 1980s, most remarkably in the work associated with Appadurai (1986). Recently, Things have returned to theory, in the form of commodities, technologies, foodstuffs, urban spaces, and art, and "the sensuous immediacy of the objects we live, work and converse with, in which we routinely place our trust, which we love and hate, which bind us as much as we bind them" (Pels, Hetherington and Vandenberghe 2002: 1). The work being signalled now indicates the emergence of a fundamental ontological confusion between human actions and the life of things, and the search for a way of comprehending this.

Over recent times, the apparently alarming qualities of the new machine-life have quickly been normalised, nowhere more so than in the way infants and children are given over to this realm of experience from earliest days. The way the infant and child experiences its object-world is fundamentally determinant of its lifetime relationship with Things.[1] In unexpected ways, perhaps, the world of machine-production has been training each generation not just to accept, want or use its Things, but to love them, to love with them, to use them for purposes of loving. This is because Things become Objects, and Objects are, in the psychic world of the infant and young child, both Thing and Not-Thing, both possessed and possessing. This makes their qualities uncanny, and the uncanny, familiar. From child to adult, the relation to the object follows familiar trajectories, as human objects and non-human objects lose their distinctiveness and begin to blur, to melt into one another and fuse. In this way, through the penetration of subjectivity by a world of Things, another kind of world (one which is already here and already constituting us) draws us as a species forward into the condition of post-human being.

THINGS OF MODERNITY: A BRIEF HISTORY

During that decisive period of European self-constitution associated with Descartes (1596–1650) the foundations of recent Western thought were instituted, resulting in a view which disassociated the mind from the body. This view suggests that only human beings have a mind, and the qualities of the world beyond the mind can be known only by the use of reason, a quality of mind alone. The mind in its substantial essence exists independent of the body. Thus Descartes was able to make the body itself into a kind of Thing, an object which must confront other objects of both similar and different kinds, none of which could be known other than by the mind. The human body itself is only another kind of object to the mind which inhabits it.

Other "things", organic matter such as rocks and trees, animals and other phenomena of the natural world, exist also only as objects for the thinking subject. And finally when we come to the things made by humans, the "technological" items with which human life is surrounded, these have even less claim on existence than organic things or human bodies. They are thought of as created by human need or desire: the mind reasons in such a way as to recognise the need for the thing and then imagines it. Subsequently methods of creating it are found. Once created, it is still just a thing, but now one which "belongs" to its creator because the latter has invested labour in producing it. This view underlies the logic of modernity.

The question of humans and their things can be regarded as an aspect of the ethics of recognition, a convergence emerging from certain streams in psychoanalysis and philosophy which leads towards questions about intersubjectivity in a post-Cartesian world. This Hegel-inspired debate centres around the idea of a concept of spirit (*Geist*) which arises from the affirmation of mutual recognition between subjects. Recognition (*Annerkennung*) underlies both the Master–Slave relation, and relations embedded in civil society. It has been argued, by Williams for example (1998), that a full intersubjective recognition in human life is the only possible basis for an ethics. In contrast, neo-Lacanian theory suggests that claims such as Williams' offer no more than a utopian hope arising from the hegemony of ego-psychology and its derivatives.[2] While questions of mutual recognition and their ethical import are almost wholly referred to the human dimension (intersubjectivity of human beings in their relationships) there is also a potential for exploring the intersubjectivity between humans and their things, recognising that humans are always and inevitably both subjects and things, for themselves and each other, while their things may become their objects and be attributed some level of subjectivity. To the extent that this can occur, the Cartesian world view loses its adherence.

Hegel wrote extensively of the human relationship with things, expanding on the logical implications of the Cartesian perspective.[3] In *The Philosophy of*

Right we find that the person is in effect separated from himself; and the thing is absolutely separated from the person. People, in turn, are dependent for their substantive existence by reference to their relationship to things. The freedom of personhood lies in the necessity of distinguishing oneself through the possession of property from another and "it is only as owners that these two persons really exist for each other" (Hegel 1967: 38).[4] The "thing" is external pure and simple, "something not free, not personal, without rights" (Hegel 1967: 40).

> A person has as his substantive end the right of putting his will into any and every thing and thereby making it his, because it has no such end in itself and derives its density and soul from his will. This is the absolute right of appropriation which man has over all 'things' (Hegel 41).

What then is the nature of the person who has this right of appropriation over all 'things'? This "I" may be an immediate individual, but it is an "I" which is not identical with itself, not exactly its own organism.

> As a person, I am myself an immediate individual; if we give further precision to this expression, it means in the first instance that I am alive in this bodily organism which is my external existence, universal in content and undivided, the real pre-condition of every further determined mode of existence. But, all the same, as person, I possess my life and my body, like other things, only in so far as my will is in them (Hegel 1967: 43).

So, in the thought mode which has created the world of late modernity, there is an "I" which is divisible from the body in which it exists. This "I" possesses freedom of will through which it can claim its "things", and exchange them with others by means of contract. In this moment of exchange, each can recognise the other who possesses as an equivalent being, another "I", an uncannily detached mode of inter-subjectivity.

FROM HEGEL TO MARX: THE AUTHORLESS THEATRE

The logic of modernity which has created the world of "things" as the domain of the commodity is commonly critiqued in its moral and social implications through Marx's concept of "commodity fetishism". The structure of the capitalist mode of production is one in which alienated objects which belong to various "I's" enter into a ceaseless cycle of exchange and magnification. The essential truth of this economy is disguised by its fetishisation (see Marx, *Capital* 1964 [Vol 1]: 76–8), that is, what appears to us to be a relation between things is in fact a system of social relations. This mystification results in people relating to things as objects as if they were separated from the people who made them, that is, from their production process. "A social ('human') relation cannot therefore be found behind 'things' in general, but only behind the thing of this capitalist relation" (Balibar 1970: 217).

Marx presented the capitalist system itself as a mechanism, a machinery, a machine, a construction (*Triebwerk, Mechanismus, Getriebe*): the terms and their attributes are identified in Althusser (1970: 192). The term *Darstellung*—a portrayal, a method of description—can also be used to indicate an expression of the existence of this machinery in its effects, a *mise en scene* of the theatre of the Real in which a theoretical drama is ceaselessly taking place. The whole of the edifice in a way rests on this assertion: that this theatre is directed by itself, it provides "its own stage, its own script, its own actors, the theatre whose spectators can, on occasion, be spectators only because they are first of all forced to be its actors" (Althusser 1970: 193). They cannot write the script, or act other than in accordance with a mise en scéne always/already provided by this ceaseless circulation of alienated "things", which are also pure abstractions. Thus the impossibility of a break or rupture within the system; "things" can only exist in this form because they have no freedom of their own, and can have none according to the logic underlying their relation to existence, that is, their Being, as existing only for the "I" which has determinate being.

HEIDEGGER: THE ENDURING (*DAS WAHREN*) OF THINGS

Heidegger moved further in explaining the basis for the contemporary relation between human life and the world of things. In *The Question Concerning Technology and Other Essays* (Heidegger 1977) he insists that the world of things is not something neutral, deriving simply from a form of instrumental human activity. But where for Marx the mistake is in supposing the "thing", that is, the product of technology, has a value derived independently of the human who makes it, Heidegger asks that we stand back entirely from an instrumental definition. While there is no doubt that "the will to mastery" underlies it insofar as the human is concerned, the possibility that something else is at work is raised. Rather than seeing the thing as the product of the human will-to-mastery, it could also be seen as something waiting to be brought into presence, waiting to be brought forth. *Poeisis* is the Greek word for bringing forth into presence: not only "things" in the form of handicrafts or objects, but also art, poetry and performance can be seen in this light. The object, and he takes the case of a silver chalice, is brought forward into being by using the craftsman or the artist. So "[t]echnology is no mere means. Technology is a way of revealing" (Heidegger 1977: 12).

As true as this might seem for the arts of cultural production, and for simple handiwork and craftwork, how can it apply to modern technology, based as it is on the laws of physics rather than the laws of *poeisis*? Modern technology throws a challenge to nature itself, asking that energy be extracted and stored, refiguring "nature" as no more than a supply of raw materials; modern technology demands that the earth reveal itself in a different way, so that "the earth now reveals itself as a coal-mining district, the soil as a mineral deposit" (Heidegger 1977: 14).

This is a mode of unconcealment which requires that everything must stand by, be immediately at hand, available on call as required by the human will. Heidegger's term is "the standing reserve". "Things" now become something else: "stock" or "stand-by". Heidegger's critique of modern technology is thoroughgoing; but he concludes that there may be a "saving power" in technology, even in spite of the dangers. His tone is one of nostalgia: for a world where technology was an aspect of *poeisis*, or art. And yet he concludes that even aside from the implication of science and in particular physics with the growth of technological ordering, it is also possible that technology is not opposed to *poeisis*: it is part of it, embedded within it, an expression of art.

> Could it be that the fine arts are called to poetic revealing? Could it be that revealing lays claim to the arts most primally, so that they for their part may expressly foster the growth of the saving power, may awaken and found anew our look into that which grants and our trust in it?…the more questioningly we ponder the essence of technology, the more mysterious the essence of art becomes (Heidegger 1977: 35).

Whether or not the saving power is to be found in *poeisis*, in art, Heidegger's perspective opens up another set of possibilities. We can suppose the silver chalice is waiting for the craftsman to produce it, to reveal it, but so too could all the other elements of technology. In other words, rather than supposing "things" to be the product of human will, possessions of that transcendental thinking subject, the "I", what if the "things" are bringing about their own invention, producing themselves by using the human will-to-power?

This brings to mind Claude Lévi-Strauss' famous statement that men do not possess myths, but myths think themselves in men's minds, and without them knowing it ("les mythes se pensent dans les hommes, et a leur insu": Claude Lévi-Strauss *Le Cru et le Cuit* 1964: 20). Myths and machines share common characteristics: they define and create cultures, perhaps even each other. Could we go further and say that many modes of thought and the technologies that express themselves through them lie quietly in wait for the human to come upon them so they can reveal themselves?

Take the case of the emergence of basic computer code, the binary model of 0 and 1. The Pakua, the symbol of the Eight Trigrams, is the central ritual symbol in Chinese Taoism. The eight trigrams are arranged according to the four cardinal and four intermediate points of direction, thus covering the eight points of the compass. Various other series of symbols may be arranged inside or outside these. The eight trigrams are paired with each other into 64 possible combinations of six lines. Among the differently arranged sequences is the Fu Hsi Sequence, which is the earliest one (Fu Hsi is known as the First World Emperor, said to have lived from 2953 to 2838 BC). For 2000 years this series of symbols was known and used for magical and divinatory purposes particularly in the southern parts of China. Not until the seventeenth century did it become possible for this system of thought to be "recognised" by the West. Missionaries travelled to China to bring Christianity. Among them was Father Joachim Bouvet, a Jesuit priest, who became fascinated with Chinese systems of orthography and representation. He showed Gottfried Wilhelm Leibniz the Fu-Hsi arrangement of hexagrams. Leibniz is known as the Father of Calculus and the first Westerner to "discover" the binary system. Perusing the various hexagram arrangements, he found that taking 0 for each solid line and 1 for each broken line, reading up from each of the hexagrams, the sequence 000000, 000001, 000010, 000011 was created. This was the system of binary code for the first 63 numbers. On this basis, binary languages could develop, without which the technology of the late twentieth and early twenty-first century would not have been possible (see Ong 1991: 64–7).[5]

PSYCHOANALYTIC OBJECTS

From *techne* to *psyche*: the idea of the "object" in psychoanalytic theory opens up a different but related perspective. In Freudian and post-Freudian thinking, the term "object" refers in the first instance to the attachment process securing the infant's connection to those on whom it relies for survival. The paradigmatic object is the breast itself. The baby is "attached" to the breast, both really and metaphorically. Without this attachment, it will die. And yet it is not securely attached, because the breast is attached to a body and the body can, according to its own will, move away or fail to appear in accordance with the desire of the child. The complex and difficult processes whereby the infant comes to understand the relation between self and other have formed the primary elements in most psychoanalytic theories, notably those of Freud himself, of Melanie Klein and of Lacan.

Psychoanalytic theory is notoriously ethnocentric. The relation between the infant and the breast is the paradigmatic relationship in the development of a sense of subjectivity and its encounters with an external reality. Freud and the post-Freudians saw many problems in this relation which they attributed to a universal human experience. In Freud's time, and frequently among upper-class people in the previous several centuries, the infant's attachment to the breast was in various ways compromised. In particular, wet-nursing may have meant that the infant was usually nourished by someone other than the mother. This often resulted in precarious nutritional status and perhaps an overall lack of care. For many other infants, though they were nourished by their own mothers, there were also various strictures regarding frequency of feeding, the need for privacy, insistence that the child sleep alone or away from the maternal body, and the cultural preference to terminate breastfeeding as soon as possible. The first "object" was almost certainly a very problematic kind of attachment under these cultural regimes. Theorists such as Melanie Klein claimed that these problems caused profound depression, anger and hatred in the very young baby. The presence of maternal aggression and hostility to the infant was also developed in Kleinian theory as a significant element of the earliest formation of universal human object-relations.[6]

But in the majority of tribal and pre-modern societies, this "detachable object" is in fact scarcely ever detached until the infant is well able to cope with the experience. The breast may be attached to the body of the mother, but so is the baby. The baby is carried physically on the mother's body wherever she goes; the mother gives the breast to the infant whenever it cries. The subjectivity of the mother and her status as part-object for the baby are not normally in conflict. Breastfeeding continues for at least two years and often up to four years. By this time the "object-status" of the mother is no longer a problem: the child has recognised itself as a subject, and knows cognitively that the mother is a separate being, even if in its fantasy–desire this were not so. This is not to imply that there are no "failures", or that the mothers have no ambivalent feelings about their infants. Babies were often killed or left to die for various reasons: if they were too close to another sibling, or one of twins, or if environmental conditions would render their survival unlikely. Nevertheless the confusion between subject and object, identity and otherness, arising from the cognitive underdevelopment of the very young infant exposed to rigid and sometimes neglectful or hostile early nurturance, would not be likely under pre-modern conditions.

However, technological developments have rendered almost completely irrelevant the fundamental nurturance bond between mother and infant, and the specific relationships with the maternal breast. Although artificial feeding of infants was tried on occasions since Ancient Egypt, only since the twentieth century has it been really effective or possible on a mass basis. Modern technology permitted the development of the feeding bottle and the rubber teat, modern manufacturing developed dried milks which mimicked the qualities of human milk, and the germ theory allowed the development of reliable sterilisation. Now, all over the developed world, the "object" which nourishes the infant is increasingly likely to be a manufactured "thing". The development of artificial feeding apparatus is now so normalised that the question of whether or not to breastfeed has become merely a lifestyle issue. Every woman in every part of every developed society on the planet can expect to go to a store and buy 100 per cent reliable artificial milk, bottles, teats, sterilising equipment, and the like. Many women breastfeed for a few months but most children are "weaned" by six months of age or earlier. They are not really "weaned", but the bottle replaces the breast and usually the "dummy" is provided for non-nutritive sucking. At an early age, most infants can take control, to a degree, of their own feeding. By seven or so months of age, or earlier, infants can take their own bottle into their own hands and feed themselves. They can continue on the bottle as long as, or longer than, pre-modern and tribal children can access the breast, and go to bed each night with a bottle of milk no matter what their mothers or carers are doing. The dummy is attached to their clothing and they can put it in and out of their mouths as often as they like for as long as they like. The "object" is in the infant's own control, and it is free to love its "thing" provided the carers provide it. Every parent can confirm that the baby really does love the bottle and the dummy. This loved "thing" does not have to be confused with the primary carer, who is free to come and go while the infant retains control over its loved object. Of course the infant also loves the human carers but its particular dependence on the maternal breast is no longer an essential part of that relationship.

Nevertheless bottle and dummy only "stand in" for maternal love at a particular point in development, and as the infant begins to recognise itself, and know that its primary carers can and do go away and leave it behind, the security of its love attachment comes into question. The conditions of modernity, which have permitted the mother to detach from the baby for the purposes of feeding, also mean that the mother may be absent from the infant, and increasingly infants are expected to accept this without "making a fuss", spending large amounts of time at day care centres from an early age. There is a debate as to whether this affects the infant's sense of attachment; whether this happens or not, we also know that no level of continuity of attachment can ever be enough. Some level of maternal love is always missing. The demand on the mother will always be thwarted: by the father or other relatives' demands, above all by the birth of new siblings.

Subsequent love-objects may then be seen as a consolation, a supplement to the missing maternal love. In this way, the person loved functions as a kind of "thing" for the person who loves, and just as for the baby the mother is a separable person and a kind of Thing, so is the love-object for the lover. In Lacanian theory, a person can never know or encounter another person in their entirety or in the fullness of their being: there are always ideas, images, fantasies which intervene, arising from the problematic confusion between subject and object. Between any two people is a space full of projections and desires arising from the "objectification" of the other, and thus the impossibility of a completely transparent intersubjectivity.

If we cannot grasp the "real" of the other when the other is a person, why do we assume we can grasp the "real" of the non-human object? Could it be that the non-human object presents itself in ways every bit as obscure and mysterious as the human one?

THE TRANSITIONAL OBJECT

As child-observers well know, children do not recognise a clear distinction between the "I" and the Thing for many years. The boundaries between them are hard to establish. Children frequently form deep attachments to particular objects—a favourite "toy", a piece of blanket—without which they feel unable to function adequately, especially as they approach that terrifying space between being alive and being dead, that is, the time of going to sleep. These are known as "transitional objects" in Winnicott's terms. As well, most children find it difficult to ascertain exactly what is "real" and what is not. The testing of "the real" goes constantly in questioning, chattering, in fears (as in "is there a monster in the cupboard?") and in the ideas of being able to do something which is desired, for instance, the child believing it is able to fly. Many children also believe that the toys and other objects which they love and share their lives with actually do have a life of their own, as in the film *Toy Story*, where the "things" in the child's room have a secret life which happens during the night, their animated existence merely being put on hold during the day. Transitional objects are particularly important for providing reassurance against fears of "the real" and the dangers and terrors faced by children in their imaginations (and all too often in reality).

At one level, the transitional object acts as a controllable supplement for the missing maternal/paternal carer. At another, though, transitional objects may be a compensation for the many objects which the child is not permitted to touch directly, explore or enjoy. Modern societies are replete with these kinds of objects, all of which must carefully be kept away from children. In pre-modern societies children are given free access to almost all objects in their environment. Howeve, there are always some objects which they must avoid: and these include those for which the term "fetish" was originally intended— sacred and secret objects which are culturally and socially believed to hold important powers in and of themselves. The objects which are most stringently protected are those which are objects of the greatest "value". In the everyday life of modern societies, the environment of the child is also composed of fetishes, in this case items of value which might be destroyed if the child were to touch them. The process of growing up then is learning to respect the inviolability of the fetish.

For adults in contemporary societies, this relationship forms the core of our relationship with our fellows. We must not take another person's object because it belongs to him or her, in its substantiality. At the same time, we can neglect or destroy "our own" objects, if we choose to do so, but it would only be the most unwise, or crazy, who would do this. All of our objects partake in the character of fetish objects. In tribal societies the logic of the fetish is quite different. The fetish-object is imbued with power of its own. People must respect the object because it possesses this power which it will direct against those who trespass against it. But objects in the everyday world need not be held aside, since they can always be replaced by the efforts of the individual. It is often observed that in many tribal communities "valuable" things such as cars, musical instruments, television sets, video players, and the like, are quickly wrecked because children are allowed to "play" with them. Children are free to explore the objects because they are not considered to be continuing assets which must be retained in their original condition. This is a logic based on the satisfaction of desire, and on a sense of abundance in the world rather than scarcity. We are willing to relinquish because we can always obtain more. Shortage is a particular kind of condition, based on a perception of objects as alienated in their essence from the desires of the human world.

LOVE AND ITS OBJECTS

Thus we can reconsider the world of material objects from the viewpoint of love. Under late modernity, the two seem to converge more and more completely. The bonds of love are redirected towards the Thing, or perhaps expressed most completely by reference to it. The love directed by parents towards children, for instance, can be subject to processes of substitution. The man, for example, may celebrate the purchase of a new car as if it were a child, albeit with a kind of ironical distantiation. The transference of love from a primary object to another was accepted by traditional psychoanalysis as a normal process. For example, the love for the mother is supposed to be echoed or rediscovered in the man's love for his partner, under the logic of Oedipal heterosexualisation.

Where a love-object has disappointed someone, this loss is turned inward and becomes melancholia. The wounded self-regard which arises from the loss of love is turned into self-reproach and depression.

> An object-choice, an attachment of the libido to a particular person, had at one time existed; then, owing to a real slight or disappointment coming from this loved person, the object-relationship was shattered. The result was not the normal one of a withdrawal of the libido from this object and a displacement of it on to a new one, but something different...the shadow of the object fell upon the ego, and the latter could henceforth be judged by a special agency, as though it were an object, the forsaken object (Freud in Peter Gay 1995: 586).

In the Real, there is always and inevitability the loss of loved object. Parents grow old and die, partners leave one another, children leave home and leave their parents. In contemporary modernity the bonds of "love"—in a form which would have been recognised by Freud, say—seem to have weakened. The pair-bond of the reproductive couple is totally optional and often an afterthought, if it happens at all. (Postcard: "Oh my God, I forgot to have children!") The mother-child bond remains the strongest and most enduring tie, but the arrangements surrounding it are changing rapidly. International adoption, surrogacy and in-vitro fertilisation (IVF) are reconfiguring the forces of social and biological reproduction.

Interpersonal relations become increasingly problematic. Where family groups were once large and expected to be enduring, increasingly co-residence is becoming rarer and people seek to, or are obliged to, live alone. A major element in this reduced sociality is a love-object which is intermediate between the person and the thing. The pet, or animal companion, provides an object which is both substitute and transitional; the pet is manageable and responsive, inexpensive and short-lived in contrast to children. But even pets have severe disadvantages arising from their existence as "things" which exist organically, disadvantages which are being addressed by SONY as a result of its huge research effort in Artificial Intelligence. SONY's new AIBO robot dog is a Second Generation entertainment robot: it learns its name, and "its built-in stereo microphone, voice-recognition technology and speaker form the foundation of a growing, loving relationship".[7] AIBO can learn about 100 verbal commands and expresses "real emotions". "AIBO has instincts and acts as if it had free will".[8]

The robot dog has been successful in Japan, less so elsewhere. But no doubt it is the forerunner to other creatures/beings which will join our human lives in an intermediate space between Being and Not-Being. Libidinal attachment and "feeling good" in the world depends on objects which may increasingly be objects rather than other human subjects. The fundamental break between the "I" and the "thing", signalled by Descartes and carried forward into the world of global technology, may be coming to an end, thus opening up vast possibilities and dangers.

THE UNCANNY AND THE UNDEAD

We can see that there is a deep element of the Uncanny in our relations with objects; tribal societies knew it and we are beginning to know it again now. The material world increasingly "stands in" as well as "standing-by": an extension and a reserve for our own desire. The commodity fetishism of late capitalism has made this its engine, its ceaseless thrumming engine. As Zizek has described it, the indestructible stupidity of superego enjoyment enfolds us with the Master's injunction: "Enjoy! (Or else!)". This is the clutch of the "undead" as it constitutes the perverse universe of late capitalism (Zizek 1999: 390):

> Obedience to the Master is thus the operator that allows you to reject or transgress everyday moral rules: all the obscene dirty things you were dreaming of, all that you had to renounce when you subordinated yourself to the traditional patriarchal symbolic Law—you are now allowed to indulge in them without punishment, exactly like the fat-free German meat which you may eat without any risk to your health...(391)

The dilemmas of drive and desire, and the ethics required in the face of them, must lie at the end of this story, but we cannot reach it yet. The dimension of the "undead" is that of "a strange, immortal, indestructible life that persists beyond death...an even worse infinity of *jouissance* which persists forever, since we can never get rid of it" (Zizek 1999: 294). If it is true that we can never get rid of this uncanny relation with objects, the Undeadness will persist. But as we begin to contemplate the possibility that the objects are themselves "alive" the estrangement between the "I" and the "thing" may find a mode of dissolution.

The logic of capitalism thus is imbued in its unconscious practices with a particular understanding of objects in the world. Technology has created objects beyond number, in an unthinkable abundance by comparison with conditions even in the recent past. Our discarded objects mock us in the garbage dumps of the world. If objects are recognised as having a kind of subject-hood, or to express it in a different way, if the strict separation between object and subject can be abandoned as much in the object-world as in the intersubjective world, a more care-full and respectful relation between humans and things might arise. New technologies open up these possibilities. Exploration of space and harmony, the mapping of human genetic code into musical form, the responsiveness of new objects—the robot dog, the speaking computer—signify the possibilities at the edges of thought, and art.

ENDNOTES

1. In a work-in-progress I am looking more closely at the relationship between infancy and mastery under modernity, and the way in which the forms of subjectivity appropriate to the machine-world are established in early life.

2. Lacan was insistent that the appropriation of psychoanalytic theory by the American establishment, and its assumption of a therapeutic practice guided by ego-psychology, represented a fundamental failure to understand the social and cultural implications of the Freudian inheritance. Today, this could be seen as a basic cultural underpinning for the dissolution of mutual comprehension between the United States and "old Europe", although this argument must await another occasion. A voluminous literature exists on Lacanian psychoanalysis and elements of early Lacanian theory have entered many areas of the humanities, but the recent emergence of a neo-Lacanian theory of the social and of subjectivity is notable. The theory of the Void in intersubjectivity and the theory of the Four Discourses imply that subjects can never fully recognise one another, and inevitably "speak" from varying positions aligned around the presence or absence of Mastery and its signifiers. With regard to the formation of the subject in "classic" Lacan, see his paper "Position of the Unconscious" published in *Reading Seminar Xl* 259 ff. (Feldstein et al eds. 1995) New York: State University of New York Press. Two accessible accounts of the Theory of the Four Discourses are found in Bracher's paper in Bracher et al(eds.) 1994, see also Nobus, 1998 for a lucid summary. Lacan's Seminar Seventeen, *L'envers de la Psychalanyse*, where the theory was most clearly articulated, has not yet been published in English and the French edition is difficult to obtain (Lacan 1991).

3. Williams (1998) argues against the view that Hegel insists on the separation of the person from himself and from the thing. I cannot see how any reading of the relevant passages in *The Philosophy of Right* could lead to that conclusion.

4. References here are to the Clarendon Press English translation by T. M. Knox, first published in 1952 and subsequently reissued as a paperback edition by Oxford University Press in 1967.

All page references are to the paperback edition. The work was first published in 1821 under the title *Naturrecht und Staatswissenschaft im Grundrisse* but was also issued with another title *Grudlinien der Philosophie des Rechts*. Reference by the translator has also been made to a number of other editions: it is almost impossible in Hegel scholarship to clearly differentiate the components used from the "original" versions due to the interpositioning of fragments of texts from different sources in different early editions.

5. While calculus and the binary system are distinct mathematical entities, Leibniz played an important role in developing both. He never found a way to combine them.

6. Klein focused on the presence of destructive impulses and fantasies stemming from oral sadism towards the mother, and the internalization of a "split" relationship between a devoured and devouring breast on the one hand, and a satisfying and helpful one on the other, as central to the construction of the superego. See, for example, Klein 1927. There are major differences between Kleinian and Lacanian interpretations. See Burgoyne and Sullivan (eds.) 1999.

7. See http://www.aibosite.com/sp/gen/index-2.html

8. Ibid.

REFERENCES

Appadurai, Arjun (1986) The *Social Life of Things* Cambridge: Cambridge University Press.

Balibar, Etienne (1970) "The basic concepts of historical materialism" in Louis Althusser and Etienne Balibar (eds.) *Reading Capital* London: New Left Books.

Bracher, Mark and Marshall Alcorn Jr. et al (eds.) (1994) *Lacanian Theory of Discourse: Subject, Structure, and Society* New York and London: New York University Press.

Burgoyne, Bernard and Mary Sullivan (eds.) (1999) *The Klein-Lacan Dialogues* New York: Other Press.

Feldstein, Richard, Bruce Fink and Maire Janus (eds.) (1995) *Reading Seminar Xl. Lacan's Four Fundamental Concepts of Psychoanalysis* New York: State University of New York Press.

Freud, Sigmund (1995) *The Freud Reader* P. Gay (ed.) London: Vintage.

Hamilton, Annette (1980) *Timeless Transformation: Women, Men and History in the Australian Western Desert* PhD thesis Sydney: University of Sydney.

Heidegger, Martin (1977) *The Question Concerning Technology and Other Essays* New York: Harper Torchbooks Paperback.

Hegel, G.W.F. (1967) *The Philosophy of Right* Oxford: Oxford University Press.

Klein, Melanie (1927) "The psychological principles of infant analysis" *International Journal of Psychoanalysis* 8: 25–37.

Lacan, Jacques (1991) *Le Séminaire, Livre XV11 L'envers de la Psychanalyse, 1969–1970* Paris: Seuil.

Lacan, Jacques (1995) [1964] "The position of the unconscious" in Richard Feldstein, Bruce Fink and Maire Janus (eds.), *Reading Seminar X1* Albany: State University of New York Press.

Lévi-Strauss, Claude (1964) *Le Cru et le Cuit* Paris: Plon (transl. *The Raw and the Cooked* London: Cape 1970).

Marx, Karl (1990) *Capital. A Critique of Political Economy, Volume 1* Harmondsworth: Penguin/ New Left Books.

Nobus, Dany (ed.) (1998) *Key Concepts of Lacanian Psychoanalysis* London: Rebus Press.

Ong, Hean-Tatt (1991) *The Chinese Pakua, an Exposé* Selangor Darul Ehson (Malaysia): Pelanduk Publications.

Pels, Dick, Kevin Hetherington and Frédéric Vandenberghe (2002) "Editors' Introduction to: The Status of the Object" Special Issue of *Theory, Culture and Society* 19/5–6 London: Sage 1–21.

Spencer, Sir Baldwin and Frank Gillen (1968) *The Native Tribes of Central Australia* New York: Dover Facsimile Edition (First published 1899).

Williams, Robert (1998) *Hegel's Ethics of Recognition* Berkeley: University of California Press.

Winnicott, D. (1953) "Transitional objects and transitional phenomena" *International Journal of Psychoanalysis* 34: 89–97.

Zizek, Slavoj (1999) *The Ticklish Subject: The Absent Centre of Political Ontology* London and New York: Verso.

[PATRICIA PRINGLE]

The spatial implications of stage magic

In this chapter I consider a body of performative work which found a mass audience in the late nineteenth century, that of the stage magician, and suggest that ideas which were embedded in that work prefigure (albeit unintentionally) thoughts, desires and fascinations which are still at play in contemporary spatial culture. These ideas nowadays are expressed through works which, at their extreme, seem to be designed not merely to perform but to perform "impossible" operations such as dematerialising, defying gravity, vanishing, leading to an "architecture of lightness [in which] buildings become intangible, structures shed their weight, and facades become unstable, dissolving into an often luminous evanescence" (Riley 1995: endpapers). This interest in spatial ambiguity, reflection, perceptual layerings and the thickening of space by light, which informs the work of contemporary "artists of light and space" such as James Turrell or Dan Graham, is also current in interior architecture. Equally, the desire that space should transform, change shape, fold in on itself or put a boundary to infinity seems to be driving some of the most interesting work of the moment.

CONJURING WITH SPACE

Why do I propose a resonance between late nineteenth-century stage magic and the emergence of an architecture of performance and experience? Later I discuss more elusive similarities, but I was first struck by the simple fact that many of the practical techniques used by the old-fashioned illusionist are second nature to a designer.

For a demonstration of some pragmatic effects, consider this tricky table which from the front is supposed to look like a delicate piece of furniture but which actually conceals a space large enough to contain an assistant.

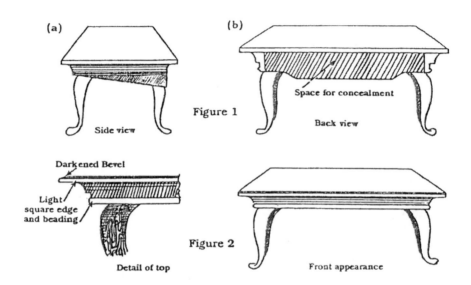

Conjurors' Tables in Sam Sharpe, *Conjurors' Optical Secrets* p.16 (1985) ©Hades Publications

The illustration is taken from *Conjurors' Optical Secrets* by Sam Sharpe (1985: 16), one of a series of volumes in which he catalogues optical, mechanical, hydraulic and psychological effects at work in magic tricks.

The table demonstrates the following (Sharpe's terminology is in italics): optical illusions of *apparent decrease by division*, such as *broken contour* and *bevelling*; other subtle illusions created by the understanding of how the eye will read dark and light surfaces; *concentric confusion* when the eye is bewildered by the multiple horizontal lines of the beading; the use of the *streamlined cavity principle* which takes advantage of the way that the eye and brain decipher perspective; the *thin edge principle* which is the same one that often makes the true depth of a television set take us by surprise; plus spatial illusions caused by people's mis-assumptions about the size of a human body, usually so much smaller than we realise. The trick would be further enhanced by illusions generated by the plasticity and elasticity of materials, and of the human body itself.

Sharpe's notes were written in the 1970s and 1980s towards the end of a career involved with documenting conjuring tricks. They represent a straightforward twentieth-century theory of the psychology of perception and most of the observations could be related to the working knowledge of a professional interior designer or industrial designer today. But as well as demonstrating a manipulation of perceptual experience, do the illusionists' tricks also reveal spatial disturbances and tensions, spatial beliefs held by the audience?

Once my thoughts had turned to magic, I was naturally intrigued by the following sentence, drawn from a quite different context (the author, Thomas Moore, is a psychotherapist): "Interior design, whether amateur or professional, is largely the practice of natural magic, an attempt to arrange life for maximum emotional and practical power" (Moore 1997: 371). "Natural magic" means the application of phenomena derived from the sciences, phenomena which have an objective existence. An example of the covert use of natural magic to maintain power would be the apparatus supposedly used by Greek temple priests which caused the temple doors to open when a fire was lit on the altar. What I find interesting here is the identification of that which we still find marvellous, magical, enchanting in the temple trick. We do not find it charming that air expands when heated or that a siphon works, but we find it magical that a door could appear to open by itself.

David Devant, one of the most famous stage magicians of the turn of the century, defined magic as "the feeling that we have seen some natural law disturbed" (Devant 1909: 8) In fact we have seen natural laws in operation, but have been misled or misdirected in our understanding of which natural laws were at work. My suggestion is that the contemporary fascination with spatial operations is part of a current willingness, even longing, for space to play with us, to respond to us, to trick, tease and disturb us, just as we, as designers, speak of playing with it. The desire seems to have been enhanced by the new spatial interests of the nineteenth century.

THE ACTIVE EYE

That the nineteenth century witnessed a "crisis" in the culture of space and time is a recurrent observation in twentieth-century scholarship. New modes of transport and communication created new spatial experiences and new spatial practices. New spaces of movement, such as the railway, revealed space opening and closing, unfolding and rearranging itself in layers of space moving past us at different speeds. Space, having first disintegrated, reconstituted itself as a dynamic and fluid entity.

Accustomed to the panoramic view, the eye learns to select details from the multitude. Perception of depth becomes more elastic, spaces are found within flatness—hence the interest in the flat space of Japanese art. New optical entertainments allow people to experience the moving image, and to learn to read depth into its flatness, while conversely learning that the depth is only an illusion and not to be feared. Henry James is able to speak of the literary technique he calls "foreshortening", the creation of an illusory depth by emphasising certain details, while allowing others to fade into the background.

The increasing flexibility and fluidity of vision are accompanied by the development of other ways of giving attention to space, leading to a perception of space as a flexible presence, capable of holding a charge. I suggest that the same increase in agility and ability which sometimes allows us to be the perceptual equivalent of multilingual or ambidextrous, could lead to the possibility of a shift in spatial awareness, whereby space can be experienced as a dynamic set of events rather than (or at the same time as) a two-dimensional or three-dimensional emptiness, and that this perception distinguishes the character of interior architecture.

Some examples which indicate this changing perception of the complex experience of interior space include the following: Richard Redgrave, addressing the Society of Art in 1850, pleaded that the utility of a carpet lay not only in its ability to cover the floor, but in its capacity to provide a visually continuous plane from which the various objects in an apartment could rise (cited in Giedion 1948: 357). His apprehension of the space-negating effect of a riotously patterned floor surface predates James Gibson's psychological theory of perception a century later—that space, rather than being perceived as a collection of objects hanging in nothing but air, can only be reliably perceived in relation to a continuous surface (Dunning 1991: 109).

When in 1881 Eliza Haweis, author of *The Art of Decoration*, hypothesises that the highly polished surfaces and reflective elements in a Louis XIV interior would have produced "a softened and indefinite effect" before which figures in the foreground would appear more clearly delineated, while "throwing the background in arrear", she notes that this motive "has never to my knowledge been pointed out before" although "in painting a picture, these calculations always enter in; and the idea is worthy of French wit" (Haweis 1977: 144–6).

Voysey, (whom I cite as an example of the new mood of design which sought to reclaim the interior from the grasp of the upholsterer in the late nineteenth century) writes about the relationship of colour and light to our perception of the height of a ceiling in terms which address the physiological and the psychological effects (cited in Gebhard 1975: 60–1). Mrs Haweis also has strong opinions about ceilings which are subtler than the decorator's convention of today which is "white to raise and dark to lower". Her words suggest both that she enjoyed her rooms by lamplight and that her seeing possessed the haptic quality:

"Ceilings should always be coloured, for a darkish ceiling throws no cold reflections down, and materially heightens the room" (Haweis 1977: 230). White walls "greatly diminish the size of a room, as a white ceiling diminishes its height… A dark wall adds size, because the eye cannot exactly measure the distance at which the wall stands; whereas in the case of a white wall, the eye calculates it to an inch" (1977: 217). She writes perceptively about the physical sensation of the eye as it reads a window in a wall, and is aware of the subtle working of the splayed reveal by which "the shock is lessened" (1977: 235–6). The awareness that our surroundings work on us by what they do to our physical senses, in addition to what they symbolise or signify, was part of a new spatial understanding.

All of the above observations have found parallel applications in great stage illusions, which leads me to my next proposition; that space had become an ambiguous and provocative entity by the late nineteenth century, a suitable subject for the play of imagination.

THE DISTURBANCE OF SPACE

To turn now to the manipulation of space in stage magic: what were some of the new visual/spatial experiences offered by the stage illusionist, and why were they so effective at that time? Although conjurors have been applying natural magic since the dawn of time, the late eighteenth and early nineteenth centuries saw the evolution of professional magicians as pure entertainers. The large-scale stage illusions which are the focus of my interest developed during the second half of the nineteenth century. As most of my source material is from Europe, I will focus on the Western experience.

Although the stage magician today can appear a sad character, a third-rate act at a sales convention, often stereotyped as ludicrous, melancholy, shabby, or somehow impotent, the nineteenth century stage magician provoked a very different reaction. In the period under discussion, magic shows became hugely popular, reached a peak of creativity around the turn of the century and had largely declined in popularity by the 1930s. Although enthusiasts continued to enthuse, the acts remained much the same but failed to interest large audiences until the recent surge of interest in the type of extravaganzas which might be described as "living special effects".

The Victorian magic shows combined traditions from the theatre, the pantomime and traditional conjuring with the new nineteenth-century amusements derived from the popular exploitation of science, such as the phantasmagorical lantern and the class of illusions known as "Pepper's Ghost". To give just one example of their popularity: John Nevil Maskelyne and his successive partners were able to stage magic shows in London continuously for more than 50 years, beginning with a 30-year run at "England's Home of Mystery", the Egyptian Hall in 1873 (Jenness 1967).

Unlike music-hall or the funfair, these stage shows contrived to be considered as suitable for the respectable, professional or educated classes, in short for an alert and modern-minded audience, and to some extent edifying insofar that the audience had been challenged intellectually as well as had been amused. The majority of well-known performers were vehement opponents of charlatanism and all mysticisms; indeed there is a tradition of performers developing their skills in order to expose the trickery of those who claimed supernatural powers. However, an atmosphere of mystery and inexplicability was the expected convention with audiences and performers both.

A distinguishing feature of magical innovation in the second part of the nineteenth century was the proliferation of apparatus whose conception required marrying the craft of the cabinet-maker and the understanding of the engineer with the designing mind of the magician/performer. New technologies might allow some tricks to be mysterious so long as the technology remained esoteric such as those involving electricity or magnetism, but in general the genius lay in the perhaps intuitive grasp that the innovative magician had on how the trick would work on the audience, and not just on how the phenomena could be manipulated. A category of tricks which came into prominence during the period is that of optical tricks involving mirrors and glass in large and complex configurations. The mirrors were able to take a piece of space from one part of the stage and make it appear to be somewhere else. The effect could be used to conceal the presence of people or objects that might be standing right in the centre of the stage, or to project a virtual image which could be taken for reality. Further categories which became important were tricks suggesting levitation, tricks using steps and stairs as "innocent" devices and tricks using light and surface to manipulate spatial perceptions.

There is less secrecy about the mechanics of the tricks than people assume. The magic lies in the art and its application, not the apparatus alone. Magicians have been publishing their techniques since the invention of the printing press. Part of the charm of the set-piece illusions at sideshows was seeing how they worked, and this would be revealed for an additional admission fee (Hopkins 1976: 60). In the famous public lectures given at the Polytechnic in 1865, the basic principles of the illusions which are done with mirrors and with projections were made public for anyone who wished to know them. Similarly, the stage technique known as "black art" where an object or performer appears invisible in front of a self-coloured backdrop is not in itself baffling, although it relies on an audience reading a deep space as flat. However, knowing the principles by which effects *might* be achieved is not at all the same as being able to say what has been actually seen, or understanding the sensations that the illusion has brought up. Further, there is a difference in the type of magic trick which could be practised by a drawing-room amateur and a stage illusionist, the latter making use of large installations and the facilities of stage equipment.

I often consider illusion not in the negative sense of the word but rather using illusion in the sense it has in the Spanish language, where it also means "hope". "I have illusions of winning the lottery". So beneath the trickery these illusions also express desires, things which fascinate, charm, enchant, things which were found spell-binding, glamorous, dazzling (all words which I have chosen deliberately because they have their origins in words which have to do with the casting of spells).

Perception is a participatory act, and the tricks of perception played by the Victorian stage magician cannot be separated from the psychological preparedness of the Victorian audience to sense them, and their susceptibility to their effects. I do not suggest that this was dominant in the consciousness of the inventors of the tricks at the time, although the period coincides with one of research into the psychology and physiology of perceptual phenomena. The veneer of science made them respectable, a sort of edification by puzzlement, yet the scenarios contain all sorts of unquestioned and disturbing undercurrents and images; violent annihilation, possession, dissolving of identity and confusion, coupled with vertiginous and thrilling spatial displacements. These seem to have been working at a deep level, and in ways that the audience might not yet have been able to articulate. It was not until stage magic was well into its decline that its historians started to theorise it in a way which, I suggest, emphasised its spatial play. Writing in 1932, Sam Sharpe, expanding on the categories of "primary feats" that David Devant had proposed 25 years earlier, defined them thus:

1. Productions *(From not being to being)*
2. Disappearances *(From being to not being)*
3. Transformations *(From being in this way to being in that)*
4. Transpositions *(From being here to being there)*
5. Natural Science Laws Disobeyed *(antigravity, magical animation, magical control, matter through matter, multi-position, restoration, invulnerability, and rapid germination)* (Sharpe 1932: 41–5).

Are these not spatially thrilling? A sixth category, Mental Phenomena, is the only one that does not challenge space. Again I suggest that a new fascination with the malleability of space was characteristic of the period, and that the tricks touched on some moment of rapture or rupture that the audience was ready to receive.

THE DREAMS OF THE AUDIENCE

It is said that each epoch dreams the one which follows it, implying that the dreams, visions and imaginings of one age are the pre-history of the next. The nineteenth century has been described as a protracted dreaming of the cinema, as all the ingredients which would come together to make cinema possible were almost forcing themselves into existence—the magic lantern and other optical toys, the development of photography, spectacles such as the diorama, the work of scientists concerned with phenomena of vision, and the development of scientific photographic equipment which, originally designed to stop motion, eventually led to the possibility of recording and simulating motion (Ceram 1965; Chanan 1980). And add to this the theory that it was in the nineteenth century that the audience learnt to be an audience, to sit in the dark in rows, to accept transformations as entertainment rather than manifestations of demonic powers, to resist the manipulative and bewildering aspects of the visually complex and to relish it instead as amusement. With hindsight, cinema seems inevitable; we learnt to come together in darkness and share a collective dream (Winston 2001).

The relationship between early cinema and stage magic is well documented. The first public screenings formed part of magic shows, and some of the first cinematographers, such as George Méliès, were also stage magicians, and the great-grandfathers in spirit of the people who produce special effects for cinema today. I suggest, however, that nineteenth-century stage magic also contains evidence of a collective *spatial* dream, a desire and ability to play with spatial concepts in order to feel a sense of spatial form, a spatial thrill, indeed a spatial eroticism. Just as the audience can be seen to have been essential to the "invention" of cinema, so the audience were essential to the success of stage illusions.

The nineteenth century has been described as having exhibited a horror of emptiness which led to an excess of ornament, the filling of even the centre of the room with "stuff" and the obsessive draping of every aperture (Giedion 1948: 342–4, 375). But the repression of space leads to the creation of myriad new spaces: between the curtain and the reveal, within the drawer, between the picture and its frame, under the piano, hiding in the pattern of the wallpaper and within the myriad of drawers and receptacles and workboxes and lockets. The proliferation of patented multi-functional furniture and household equipment in the period speaks of a heightened consciousness of the way in which each object has spaces and purposes latent within it.

The Victorian phobia that any space might conceal an object or person that has the potential to affect our lives is evident in fiction and imagery. In spite of, or perhaps in reaction to, improved lighting, the fear increases that darkness may rebound on us, and the home is full of dark spaces. A study of the home life of the American family from 1750 to 1870 notes that court records and early American fiction are filled with references to the unperceived "telltale" witness in the corner (Garrett 1990: 150). The insecurity of secure spaces is also a theme of fiction, perhaps best summed up by the twentieth-century novels of Ivy Compton Burnett which so tellingly work up these literary traditions in their fictionalised

High Victorian settings, where the incriminating letter often springs out of the secret drawer of the desk and lays bare the moral disarray of the household. Safes and locks were the focus of exciting thoughts leading to a variety of new apparatus from the inventor and new routines from the illusionist. The shifty nature of space is both attractive and terrifying, and just the material for entertainment.

Emptiness and potentiality are relished as disturbing sensations of the Victorian period—the vacant chair, the empty hearth are metaphors in popular art and song. The gate of the enclosure surrounding a family burial plot should not close fully but remain ever slightly open, a reminder that removal to the other world may come at any moment. Again, the veiled and the screened have a deep resonance in the Victorian mind with the symbolism of the draped urn on the grave of a child and the half-draped urn on that of a young person. Even the clothes of the late nineteenth century make extensive use of filmy layers of sheer and semitransparent voile and muslin. Hints of hidden space are all around. With this in mind, the potency of a routine like "The Vanishing Lady" can be imagined.

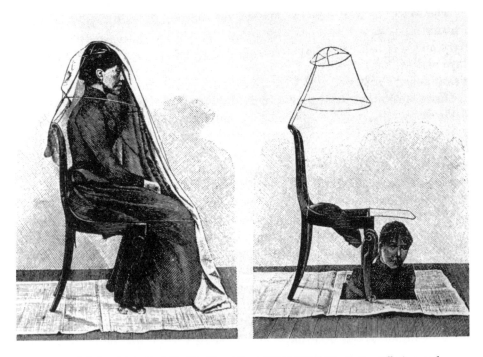

Bautier de Kolta's "Vanishing Lady" in Hopkins, Albert (1976) [1898] *Magic: Stage Illusions and Scientific Diversions Including Trick Photography* New York: Munn (reprint: Dover).

For further evidence that the second half of the nineteenth century introduced a heightened sense of enchanted spatial manipulation to the conjuror's routine, compare one collection of tricks from 1803 with another from 1898. The first book, *The Conjurer's Repository* is a mixture of chemical formulae, physics and some superstition. Tricks take the form of sleight-of-hand conjuring. The only mirror trick is one which describes how to arrange reflective surfaces so that the person who looks into a mirror will see the face of another (1803: 70).

The second book, Albert Hopkins' well-known compendium, *Magic: Stage Illusions, Special Effects and Trick Photography*, published in 1898, contains descriptions of many optical tricks which use mirrors to swap one volume of space with another; an apparently three-dimensional manipulation. These tricks, although their principles had been known for centuries (most of them are listed in the sixteenth-century Baptista Porta's *Natural Magick*) were re-investigated in the second part of the nineteenth century, first as optical curiosities such as "Pepper's Ghost", and later as stage apparatus, such as Maskelyne and Cooke's Protean Cabinet, via which one person could apparently turn into another.

In addition to the optical tricks, Hopkins describes much apparatus of the type whose trickery uses hinges, folds, springs, concealed cavities, trapdoors, pulleys and similar mechanical ingenuity. This passion for ingenious mechanical design, allowing adaptability or reversibility to challenge space, is captured in Siegfried Giedion's chronicling of the proliferation of patent furniture in the century (Giedion 1948). The germ of ergonomic efficiency is here also, many tricks relying on the performers and their assistants making strategic use of body movements choreographed in the light of precise study of the sequence of events of the trick. Space is saved, stretched, hidden, consumed, charged with expectation or made inconspicuous. If space in magic in the nineteenth century was increasingly plastic and active, what was its relationship to those who experienced and enjoyed it?

A striking difference between the two volumes is in their tone. *The Conjuror's Repository* lists phenomena dispassionately. But to look through the pages of Hopkins' *Magic* is to feel the presence of a nineteenth-century sensibility. The descriptions are rational, didactic, scientific yet somehow dreamlike. Elaborate sections through both performers and apparatus reveal the hidden spaces, the "before" and "after" of the trick, the offstage darknesses, the secret passages, the impassive expressions of the performer and his assistant, and yet suppress the contortions, exertions and physicalities of the act. Often the illustrations show an impossible view, or one which contains simultaneous viewpoints as if the observer's eye could be freed from the body and floating both in the auditorium and in the wings. The frisson is of being in two places at once, and of seeing in many different ways at one time.

Here, for me, is the clue to the thrill of some of the acts, not a tedious struggle to work out how the trick was done, but a sensuous submission to the visceral sensations induced by seeing in "impossible" ways, and experiencing space in contradictory ways in rapid succession. The conflict of vision, the rapid interplay between space sensed as deep and then as flat, as static then dynamic, clear then thickened, would produce an overload of scopic sensations, an oscillation between Cartesian and Baroque states of vision, to use Martin Jay's terms (1988). This oscillation would be felt as dizzying and disturbing, impossible to bear without the relief of laughter. The sensation is heightened by a sense of mystery and the interplay of clarity and obscurity. Noticeably, the human body is implicit in the new set-piece tricks, as the subject or the observer or the performer. And the body, and particularly the female body, is made to vanish, to lose its head, to be doubled, to be in two places at once, to be torn apart, to have its understanding shattered and remade, to be consumed by darkness and transformed in light.

Where does the disappeared object or person go? The tricks show the development of the subliminal awareness of an "elsewhere" that was not supernatural. Literally we know it to have been the space below the stage, or outside the enchanted sanctuary of the limelight. But metaphorically a previously inconceivable darkness has now been recast as a place to which people may go and return without having been visibly shaken by the experience (unlike present-day stories of being taken by aliens). Permission is given to play with "other" space without invoking madness: literally to play, for play is a serious business. Surrealism would address the psychic profundity of play. Hopkins' illustration of the *Vanishing Lady* could join the picture-novel images of Max Ernst (whose *La Femme 100 Têtes* phonetically gives us "La femme sans tête") which he acknowledged as "reminiscences of his first books, a resurgence of childhood memories" (Giedion 1948: 362–3).

Roger Caillois' theory of play (1961) suggests that when vertigo and simulation are sublimated in games and fiction, cultures are emerging from a stage of "chaotic originality":

May it be asserted that the transition to civilisation as such implies the gradual elimination of the primacy of *ilinx* and *mimicry* in combination, and the substitution and predominance of the *agôn-alea* pairing of competition and chance? Whether it be cause or effect, each time that an advanced culture succeeds in emerging from the chaotic original, a palpable repression of the powers of vertigo and simulation is verified. They lose their traditional dominance, are pushed to the periphery of public life, reduced to roles that become more and more modern and intermittent, if not clandestine and guilty, or are relegated to the limited and regulated domain of games and fiction where they afford men the same eternal satisfactions, but in sublimated form, serving merely as an escape from boredom or work and entailing neither madness nor delirium (1961: 97).

Caillois defines *mimicry* as "playing a part" (1960: 20), while *ilinx* includes those games which are based on the "pursuit of vertigo and which consist of an attempt to momentarily destroy the stability of perception and inflict a kind of voluptuous panic upon an otherwise lucid mind" (1961: 22–3) Perhaps the relish for playful manipulation of space expressed here reflects the nineteenth-century's deep reorganisation of more profound spatial concepts.

TWENTY-FIRST CENTURY SPATIAL IMAGININGS

Today magic no longer works where it did. Anticlimax and a sense of futility had attached themself to stage magic by the middle years of the twentieth century. Nowadays there is something literally exhausting about watching a dull magic act as device after device is brought on stage and its hidden spaces are emptied out. The expediency and opportunism of the apparatus becomes irritating, as ultimately facile as funny putty. Most stage magic has ceased to resonate with the spatial imaginings and desires of its period, becoming merely an anachronistic re-formatting of traditional routines whose outcomes seem inconsequential. Card tricks and prestidigitation predominate in the close-up magic which is popular in venues like hotel nightclubs, but are seen as displays of physical dexterity and clever thinking, perhaps in line with the current dominance of sport and competition.

Spatial magic has moved to other media, and is more likely to be seen in the context of an art gallery or a theoretically based design project rather than as entertainment. The "impossible" operations which gave the grand illusions their rapturous spatial thrill now form the spatial desires of much contemporary architecture. However, they no longer feel uncanny, vertiginous and likely to put us beside ourselves with shock or laughter. The sensation is more one of setting us outside of time altogether.

REFERENCES

Barnouw, Erik (1981) *The Magician and the Cinema* New York: Oxford University Press.

Caillois, Roger (1961) *Man, Play and Games* (transl. Meyer Barash) New York: Macmillan.

Ceram, C.W. (1965) *Archaeology of the Cinema* London: Thames & Hudson.

Chanan, Michael (1980) *The Dream That Kicks, the Prehistory and Early Years of Cinema in Britain* London: Routledge & Kegan Paul.

The Conjuror's Repository (facsimile of 1803 edition) Omaha: Modern Litho.

Devant, David (1909) *Magic Made Easy* London: Cassell & Co. Ltd.

Dunning, William V. (1991) *Changing Images of Pictorial Space* Syracuse University Press.

Garrett, Elisabeth D. (1990) *At Home: the American Family 1750–1870* New York: H. N. Abrams.

Giedion, Siegfried (1948) *Mechanisation Takes Command* New York: Oxford University Press.

Haweis, M.E. (1977) *The Art of Decoration* New York: Garland.

Hopkins, Albert (1976) [1898] *Magic: Stage Illusions and Scientific Diversions Including Trick Photography* New York: Munn (reprint: Dover).

Jay, Martin (1988) "Scopic Regimes of Modernity" in *Vision and Visuality* Hal Foster (ed.) Seattle: Bay Press.

Jenness, George A. (1967) *Maskelyne & Cooke; Egyptian Hall, London 1873–1904* Enfield, Middlesex.

Moore, Thomas (1997) *The Re-enchantment of Everyday Life* New York: HarperCollins.

Porta, Baptista (1584) *Natural Magick,* internet transcript of the 1658 edition at http://members.tscnet.com/pages/omard1/portat5.html

Riley, Terry (1995) *Light Construction* New York: Museum of Modern Art.

Sharpe, Sam (1932) *Neomagic: The Art of the Conjuror* London: G. Johnston.

------------- (1985) *Conjurors' Optical Secrets* Calgary: M. Hades International.

Voysey, Charles (1975) "The English Home" in David Gebhard *Charles F.A. Voysey Architect* Los Angeles: Hennessey and Ingall.

Winston, Brian (2001) "Fumbling in the Dark" *Sight & Sound* 11/ 3 (February 2001).

part TWO

[GHOSTS AND THEIR MACHINES]

[ANNE CRANNY-FRANCIS]

The ghost in/on the machine: magic, technology and the "modest witness"

In her essay, "Modest_Witness@Second_ Millennium" Donna Haraway writes about the "modest witness", the scientific observer whose disinterested observation of phenomena is central to the scientific method. Haraway deconstructs the meaning of "modesty" in this context and then situates the practice of the "modest witness" socially and culturally:

> This self-invisibility is the specifically modern, European, masculine, scientific form of the virtue of modesty. This is the form of modesty that pays off its practitioners in the coin of epistemological and social power. This kind of modesty is one of the founding virtues of what we call modernity. This is the virtue that guarantees that the modest witness is the legitimate and authorized ventriloquist for the object world, adding nothing from his mere opinions, from his biasing embodiment (Haraway 1997: 23–4).

This essay explores the relationship between magic and technology through the notion of the modest witness, the guarantor of scientific validity. Science is read here as the territorialising discourse that has worked to bring technology under the control of governments and dominant ideologies, whether they be capitalist or state socialist. Effectively, this notion of science spans what Ihde describes as "Early Modern Science" and contemporary "Big Science" or in Latour's terms, "technoscience" (Ihde 1993). Ihde notes:

> Early Modern Science…is doubly linked to technology—first in the close association of engineering and instrumentation found to be so natural in the Renaissance, and, again, in the transformation of the ideals of knowledge into experimental, interventionist forms of power/knowledge in its Baconian sense which is, again, instrument-embedded (1993: 28).

So, through the elaboration of experimentally based scientific method the practice we know as science was intimately related to technology. At the same time, science determined whether a practice was recognised officially as a technology, since technology was the practice and/or instrumentality which was linked to the development of science. By contrast, consider Don Ihde's definition of technology: "some artifact or set of artifacts—material culture—related to a context of human action or praxes (which include techniques of use)" (Ihde 1993: 32). On this basis many of the practices performed primarily by women in the home are technologies; yet they are not widely thought of as such. Instead technologies are associated with material practices performed outside the home, mostly for commercial or research purposes. The reason for this seems to be that science is associated with the latter, and not commonly with the former. That is, the scientific basis of commercial and scientific material practices is widely discussed and acknowledged; the scientific basis of domestic technologies is not. As a result science comes to define a technology as a technology in the public mind and in relation to the public purse.

In this gate-keeping role science has been a critical ideological tool, determining what can be regarded as a technology and, via the rhetoric of progress it employs, also determining which users of technology can be regarded as civilised or legitimate in their material practices. The land-management practices of Indigenous peoples of Australia were for centuries not recognised as technologies because their scientific basis was not understood. In other words the scientific discourse that prevailed at the time of evaluation was not able to situate that practice within its understanding of land-management technologies. As a result the practices were seen as incomprehensible or even as destructive (when they were "seen" at all) and the managers themselves as irrational, uncivilised, ignorant.

Here the connection between science and its definition of the rational can be seen most clearly. Science constitutes itself as a rational practice, via its elaboration of the practice known as the "scientific method". This practice of observation, theorisation, experimentation, re-theorisation, repetition, and conclusion is established as the basis of most everyday scientific practice and equated with the meaning of science itself. Of course, it might be argued that this rationalist model of science no longer applies in a post-Einstein world—that "relativity" described more than a specific theory rather ushering in a world that was no longer satisfied with the absolutist statements of Enlightenment thinking. Yet for most scientists everyday scientific practice is still governed by the discipline of the scientific method.

In a sense science validates itself through its argument for the universal applicability and validity of this method. It is this method that scientists argue takes science from the realm of the ideological, the non-rational, the irrational, faith, belief. The scientific method is the guarantor that what the scientist deals in is fact; is objective, neutral. The technologies it acknowledges are equally "valid"; they too are constituted as factual, neutral, objective. The practice of science is claimed as rational. But perhaps the greatest test of its rationality is how it deals with the non-rational.

THE GHOST ON THE BIKE

The relationship between science and the rational is explored through an encounter between a scientist, one with a vested interest in the role of "modest witness" and a ghost—a phenomenon not currently explainable within the realm of what science designates "the rational". In fact the strongest argument against acceptance of the phenomenon as rational was the refusal of the scientific observer to acknowledge its presence verbally.

The scientist in this story was at the time a doctoral student in Biological Sciences. She was in the process of writing up her research findings and was scrupulously aware of the methodology of her discipline and of the (scientific) field in which she operated. Late one night, at the time of the encounter, she was being driven by another student, not a scientist, along a main road (but not a motorway) through the English county of Essex. The driver reported the story.

On a particular stretch of road the driver became aware of a motorbike travelling behind the car. The bike was visible at a distance as a large single headlight. Suddenly the bike roared up behind the car, seeming as if it was going to ram into the back. At this point the silhouette of a figure on a large bike became visible and the roaring of a powerful bike engine could be heard. The bike then dropped back again, until it was visible only as a large headlight. It then went through the procedure repeatedly, driving very fast until it was (too) close behind the car. The driver became uneasy and told her scientist companion about the biker's behaviour so she, too, looked through the back window at the bike roaring up to the car.

Increasingly disturbed by the situation, the driver asked her companion to wind down her window and yell at the biker to stop. When the window was wound down, there was no sound and when the scientist looked back behind the car, the road was empty. However, when she, and the driver, looked through the back window, the bike and its rider were visible and, when the window was again wound up, the roaring of the bike could be heard. The driver then drove very fast until the bike dropped back and was eventually no longer visible. Since a powerful bike would have been much faster than an ancient Volkswagen, this was presumably the "choice" of the bike rider. However, the driver reported that she did not consider this at the time; her only concern was to get away from that place as quickly as possible.

The most interesting feature of this situation was the subsequent discussion. As the driver had seen several "ghosts" before, she reported that she was not disturbed by the encounter in any long-lasting sense, but found the immediate situation unpleasant. However, her companion had always expressed an implacable disbelief in the phenomenon she had described. The interesting question was, then, how would she, the scientist, deal with the situation. According to the driver, her scientist friend immediately re-confirmed her rejection of any such phenomenon. She claimed that, as a scientist, she simply did not believe in anything which cannot be explained scientifically, rationally, logically. The driver then suggested that perhaps science simply does not know yet how to describe such phenomenon. The scientist's intriguing response was: no, I simply do not accept what I saw; I did not see it.

The driver reported that her surprise at this statement was based on several factors. Firstly, it denied the reality of a shared encounter. Second, it meant that her companion had the ability to edit her reality in a most blatant manner, if not in terms of what she saw (as shared during the encounter), at least in terms of what she was prepared subsequently to acknowledge she saw. Third, it contradicted her understanding of the scientific method, which was that its first step was observation—and that this observation should be carried out as a neutral, disinterested observer. And finally, it implicitly constructed her own reality as invalid. Threaded through this list of concerns is another ghostly presence—the modest witness.

MODEST WITNESS, NOT

This essay began by quoting Donna Haraway's description of the modest witness, as a gendered and culturally specific concept that offers "epistemological and social power" to those who embody it. For the professional scientist that epistemological power resides in her or his recognition as a scientific researcher; social power in the ability to gain public funding for research and to influence public policy. In order to wield this power the scientist must be recognised as a "modest witness".

When the scientist declared that she had not seen what she had previously acknowledged, the driver was surprised by what she interpreted as a lack of scientific rigour. According to the driver's everyday understanding of the scientific method, the scientist should have followed a sequence of steps: record the phenomenon; reflect upon it and develop a hypothesis about it; find a way to test that hypothesis; develop a theory to explain the phenomenon. Ironically, it might be suggested that the driver's notion of scientific rigour was based on an acceptance of the ideology of science itself that declares its practitioner to be neutral, objective and disinterested. The working scientist knew better. She knew that to be accounted a modest witness, she could not afford to be known as someone who believes in things that are outside mainstream scientific thought. If she is positioned outside that mainstream, she will not be acceptable as a modest witness—and so will be unacceptable as a scientist. To declare herself witness to a paranormal phenomenon would be to risk delegation to *The X-Files* and the fate of Fox "Spooky" Mulder; seen as brilliant but misguided, laughed at for his preparedness to witness the "unwitnessable"—the phenomena that contemporary science cannot explain.

In this encounter with the paranormal the scientist's response might be seen as the making of a modest witness. This is most clearly seen in the scientist's readiness to edit her reality—if not in the moment of experience, certainly afterwards in the telling of the event. The driver reported her surprise at what she saw as lack of rationality in the scientist; a refusal to accept a phenomenon because it did not fit within current understanding of the real. Yet it might be suggested that, again, her surprise was naïve, as it failed to acknowledge the ideological context within which the scientist operates. If the scientist's response is looked at a little differently, we might instead suggest that in refusing to acknowledge that she had seen a ghost the scientist did not not-see it; she simply *said* that she did not see it. In other words, the focus is not on the experience itself, but on her willingness to *witness* the experience.

Haraway writes in her article on the "modest witness" about a study of the chemist, Robert Boyle by Shapin and Schaffer. The authors of the study note that three different technologies come into play when a new life form (here, the scientific method) is developed: a material technology, a literary technology, and a social technology. The literary technology is the means by which the scientific experiment is conveyed to those who were not direct witnesses. Haraway notes several implications of this technology. Referring to Shapin's later description of Boyle's experiment with the airpump, Haraway quotes his analysis of the role of the workingmen who actually worked the bellows in Boyle's laboratory: "'As a free-acting gentleman, [Boyle] was the author of their work. He spoke for them and transformed their labor into his truth'" (Shapin 1994: 406). And Haraway comments: "Unmasking this kind of credible, unified authorship of the labor required to produce a fact showed the possibility of a rival account of the matter of fact itself…" (Haraway 1997: 26). In other words, the literary account of the experiment elides the perspectives of all except the author, whose singular account is constructed as factual. Boyle did not consult his workmen; their views on the experiment are constituted as irrelevant, with only Boyle considered sufficiently disinterested to make the factual observations. So the literary account of the observation—the *witnessing*—is also part of the construction of the modest witness, and not vice versa—as the apprentice scientist understood very clearly.

Haraway also notes the literary technology's role in informing all those who could not attend an experiment:

> …those actually physically present at a demonstration could never be as numerous as those virtually present by means of the presentation of the demonstration through the literary device of the written report. Thus, the rhetoric of the modest witness, the

"naked way" of writing, unadorned, factual, compelling, was crafted. Only through such naked writing could the facts shine through, unclouded by the flourishes of any human author. Both the facts and the witnesses inhabit the privileged zones of "objective" reality through a powerful writing technology (Haraway 1997: 26).

Haraway's own writing is intriguing here in that the "unadorned" and "factual" way of writing, identified as scientific discourse, is also identified as "compelling". Compulsion would seem to be contradictory to the strictly unemotional, disembodied tone of the scientific text—a writing apparently shorn of affect. Haraway's prose, however, suggests that the very adoption of this stance carries an emotional or affectual charge—perhaps not unlike the attractiveness of the ascetic. This "naked writing" gives the reader access to a privileged realm where she or he rubs shoulders with other (first-order and second-order) witnesses and confirms their mutual occupation of a socially sanctioned and validated space. They form an in-group, a privileged subculture, a cult—with the same affective practices (reinforcement, patronage, flirtation, seduction) as other cults. To be a member one must walk the walk and talk the talk: membership is both embodied and discursive.

When non-witnesses read scientific discourse, they enter vicariously (virtually) into this space; become part of this subculture. Their own embodiment is shaped by their membership, as is their discourse. To refuse to comply is to risk rejection and ejection. The scientist who consults issues of embodiment cannot be seen as a dedicated (that is, real, true) scientist. Such a scientist might question the purpose of her or his research; be distracted by its embodied practice and so be diverted from a logical, factual, rational solution. That is, if disembodied practice is seen as logical, rational, factual, as it has so often been in modern times. As Haraway notes:

This separation of expert knowledge from mere opinion as the legitimating knowledge for ways of life, without appeal to transcendent authority or to abstract certainty of any kind, is a founding gesture of what we call modernity. It is the founding gesture of the separation of the technical and the political (Haraway 1997: 24).

It is the stance that enables the nuclear physicist working on weapons research to describe the work as "an interesting problem in physics" and to eschew any responsibility for the uses to which the research may be put—as if the research is conducted without reference to embodiment or materiality. The dedicated scientist must construct himself or herself as not only disembodied but must also communicate through a kind of disembodied prose—a prose apparently without affect which constitutes him or her as a modest witness.

In stating her refusal to acknowledge publicly that she saw what she saw, the scientist shaped her discourse to fit the acceptable parameters of contemporary science, which do not allow for (the observation of) paranormal phenomena. In other words, she constructed her account of the phenomenon by reference to, and within, scientific discourse. Her (ghostly) experience as an embodied contemporary social subject is not acceptable in the realm of science she inhabits—and so she acted as Scully to her own Mulder; she edited. But note that she edited for consumption; she was not prepared to acknowledge what she saw. Which is to say that she was not prepared to inscribe her experience within the literary technology of science, even if she privately acknowledged that experience. This was a political decision—it was also a scientific decision—and so one effectively deconstructing the ideology of science, as articulated in the concept of the "modest witness".

The driver reported that she found this practice—the refusal to witness the unwitnessable—irrational, as it seems to foreclose the possibility of a scientist discovering anything new. That is, if any event or phenomenon that falls outside the parameters of contemporary science is either disavowed or transformed so that it fits within contemporary parameters, how, she argued, could change occur. Philosophers of science such as Thomas Kuhn note precisely this problem within the practice of science; hence Kuhn's thesis that scientific advance takes place by revolutionary changes of paradigm—precisely because of the inertial drag of this kind of editing (Kuhn 1962).

IMMODESTY: A CULTURAL ANALYSIS

There is another side to this case study, however—the "immodest witness" constituted by the driver of the car. Without the demands of science to delineate her account she was able to describe her (their?) experience—to witness—but then found that her witness was regarded as scientifically invalid or unacceptable. The driver reported that, since her own experience included other such events, she had a context for situating the phenomenon, if not a scientific explanation for it. In exploring the scientist's basis for the denial of her (and, originally, their) version of the encounter she found that her own earlier experiences of such phenomenon were regarded as a reason for excluding her from "modest witness" status. That is, the fact that she had already witnessed the unwitnessable confirmed that fact that she was an unacceptable witness. She reported that, on the one hand, this argument seemed so silly that it was laughable yet it employed a denial of her experience that seemed iniquitous. It not only excluded her from those who were deemed suitably modest (with the social power and authority they wielded) but it also denied her and their common humanity—in that a shared experience was denied and her experience discredited.

The driver also reported that an ethnic dimension was added to the debate when she invoked a family history of paranormal abilities. All of the family members who possessed (or at least witnessed to their possession of) these abilities were Irish. For the English scientist, it was not surprising to find such irrational and invalid experiences associated with a culture that had, for centuries, been constructed by the English as marginal. So the driver found her credibility undermined not only in terms of her own experience (her previous history of immodest witnessing) but also culturally. Which brings the discussion back to Haraway's exploration of the "modest witness". In fact, Haraway specifically addresses the formation of Englishness in her paper, and she notes:

> Gender and race never existed separately and never were about preformed subjects endowed with funny genitals and curious colors. Race and gender are about entwined, barely analytically separable, highly protean, *relational* categories. Race, class, sexual, and gender formations (not essences) were, from the start, dangerous and rickety machines for guarding the chief fictions and powers of English civil manhood. To be unmanly is to be uncivil, to be dark is to be unruly: [t]hose metaphors have mattered enormously in the constitution of what may count as knowledge (Haraway 1997: 30).

As Stuart Hall and others have noted (Hall 1992; Michie 1992; Cranny-Francis 1995), the Irish have been constituted consistently by the English as dark, unruly and uncivil. The result of this identification is that their status as witnesses can be negated. Unlike the middle-aged, middle-class Englishman, the Irish are deemed not capable of the detachment and neutrality, the objectivity, required of the modest witness. Which does not mean that individual Irish citizens are necessarily excluded from the realm of science but rather that an individual scientist (modest witness) may invoke stereotypical responses to a specific group or culture in her or his evaluation of its suitability as scientific witness.

Haraway also notes that the "modest witness" is gendered:

> ...modest men were to be self-invisible, transparent, so that their reports would not be polluted by the body. Only in that way could they give credibility to their descriptions of other bodies and minimize critical attention to their own. This is a crucial epistemological move in the grounding of several centuries of race, sex, and class discourses as objective scientific reports (Haraway 1997: 32).

She goes on to note that these "scientific" texts about race, sex, and class constructed the body of the (white, male, middle-class) observer as transparent, invisible. Many female scientists have reported the effects of this gendering in their lives as working scientists, in the negative responses of some male scientists to their presence in laboratories and to the authority of their work. Haraway's work is valuable here in that it situates the reasons for these responses not only in the attitudes of those individual men but also relates it systemically to the discursive practice of science. In her analysis the modest witness has no body, unlike the unreliable, embodied "others" he studied—women, non-English, non-middle-class, non-heterosexual, non-middle-aged. Each of these studied categories is, through the practice of observation, em-bodied. So, like women, the non-English, non-middle-class, non-heterosexual, and non-middle-aged are "seen" to be embodied and that embodiment is constituted as the grounds on which their observations are necessarily biased; they are unreliable, immodest, witnesses. Which is to say, they cannot witness, in the terms defined by scientific discourse.

SUBALTERN PRACTICES AND MULTIPLE KNOWLEDGES

Situating the "ghost seeing" and "second sight" of the immodest Irish witness, however, suggests that this apparently non-rational, or irrational, ability may have significances beyond the everyday, as may the self-reflexive "seeing" of women and non-heterosexuals, the seers of non-English cultures, the vaunted "innocent" seeing of children and the differently-abled, and the experienced vision of the elderly. In each case, a kind of seeing is associated with those excluded from the social power and authority granted to the witness. Often this is a method of stereotyping those groups; yet it may also be the source of a subaltern response to the exclusionary power of the witness.

In the case of the Irish, constituted by the English as non-rational and so likely to think of themselves as seeing ghosts or seeing across time and space, these attributes can be a defence against the ruthless force of English rationalism. From this perspective the Irish can be seen as having been marginalised socially and culturally by the English; however, they also have magical powers that the rationalist (the scientist) cannot explain (away). Those powers mark the limits of scientific discourse, a specific inflection of the rational, of modernity (Latour 1993: 5). So magic delimits the contours of the rational, and of science, revealing it as not a disinterested, objective practice, but as a political practice—a colonising discourse that works to position a specific subject position as dominant (that of the middle-class, middle-aged Englishman). This subject position is that of the modest witness. The magic works reflexively to reveal that positioning and to offer the marginalised subject a position from which the colonising of rational, and scientific, discourse can be experienced as just that—as a discourse, not as "objective reality".

This position is not unlike that described by Teresa de Lauretis in *Technologies of Gender: Essays on Theory, Film and Fiction* (de Lauretis 1987). She notes that women experience their own construction as Woman, the essentialist category of patriarchal discourse ("equal and opposite to Man"). However, they also experience the everyday reality of being women, which means being treated as unequal. Since they constantly move between the essentialist representation (Woman) and its material reality (women), they "see" the contours of the discourse in/by which they are delimited. Women reflexively "see" their social positioning, even as they are excluded from the role of witnessing. Paradoxically, those who are permitted to witness (middle-class, middle-aged Englishmen) are blind to their positioning; they have a vested interest in not acknowledging the (located) position from which they see and speak (witness).

As Latour notes, however, the separation of critical practices (epistemology, the social sciences, the textual sciences) that once defined modernity is now under challenge (Latour 1993: 5). As these separate spheres/practices dissolve, so do their gate-keeping discourses. Other ways of thinking, of "seeing", are increasingly acceptable. In fact, they are regarded as essential if the world is to recover from the degradation of the natural and social environments caused by both capitalism and state socialism—both of which accepted the (apparent) separation of science and politics fundamental to modernity. Postmodernity is often characterised by/ as the proliferation of knowledges, with "knowledge" as a transcendent category now under challenge. Instead, knowledge (including science) is understood as a political practice, as located within a specific society at a particular time and as fulfilling particular needs (notably the construction/validation of the "modest witness"). In other words, "science" is located as a situated knowledge, not as knowledge *per se.*

It might be argued that this same situatedness has informed scientific theorisation throughout the twentieth-century. As noted earlier, Einstein's "relativity" theory introduced the position of the observer into the experimental situation. However, this observation is still qualified as this "relativism" or situatedness is commonly confined to a decontextualised, non-situated understanding of the experiment. In other words, the scientific "relativity" does not account for the relationship between the scientific endeavour itself and the society in which it takes place.

SCIENCE AND TECHNOLOGY: TERRITORIALISING DISCOURSE VS EMBODIED PRACTICE

Perhaps the apocalyptic moment of modern science, technoscience, is Oppenheimer's embodied response to the detonation of the atomic bomb over Hiroshima. He is said to have whispered a quote from the *Bhagavad-Gita:* "I am become death, the destroyer of worlds" (White 2002: 222). In this extraordinary concatenation of the poetic and the technological we see the essence of science—an embodied, situated, social, political, ethical, economic practice. When Heidegger writes of technology that it is a bringing-forth, a way of living in the world, he makes a similar point. Modern technology is, for Heidegger, a challenging-forth, which can be a ruthless drive for control of objects and people in the name of a colonising discourse. Arguably, science has operated as this discourse. But, like Heidegger's challenging-forth, the technology at the heart of science (fundamental to the scientific method) may also be the means by which it moves beyond this exploitative practice to a revealing that challenges that ruthless (inhuman, disembodied, "objective") practice. In recognising the technology at its heart, science may be led to a self-reflexiveness that acknowledges its role in modernity—the analysis we see already in the work of Shapin and Schaffer,

Haraway, Latour, Ihde and others. That is, science may be forced to recognise its embodied, material practice thereby breaking down the science/politics dichotomy that has characterised modern science. From this position scientific discourse must account for itself as a literary and social practice. Nuclear weapons research could no longer be "an interesting problem in physics"; instead its role in constituting the modern nuclear state, with all its paranoia and secrecy, would be recognised—by the scientists themselves, not just their textual and social critics.

In the meantime, it must be said, much contemporary science maintains the exclusive discourse that has characterised it throughout modernity. It is, therefore, not surprising to find that educators in the United Kingdom are still attempting to deal with the fact that there are so few non-white British scientists (*New Scientist* 9 March 2002). Despite attempts to encourage non-white British children to take science subjects at school and university, the educators have found the numbers dwindling. The reason seems to be that there is no place in the discourse of science for these students; they cannot be modest witnesses. For non-white scientist Elizabeth Rasekoala, the reason for this has to do with the value of science and technology in today's world:

> ...as societies develop, technology and science become powerful tools by which you can socially engineer the rise or exclusion of certain groups of people. In western contemporary societies maths, science and technology have all been used to allow access to some people and to keep others out (Rasekoala 2002: 46).

And she specifies the mechanism of this exclusion as the validation of only certain social subjects as modest witnesses:

> Maths has been constructed as if it's for "clever" people, and we've all bought that. And the same society that says that maths is for clever people also shows you images of what those clever people look like. For the most part, it's a white male (Rasekoala 2002: 46).

ENVOI

Some time later, in another discussion about ghosts, the scientist told her friend that her dead uncle was often seen by members of the family. Her sister had had her bedclothes adjusted one night by her uncle, apparently in a repeat of the gesture by which he had bedded down the horses that once lived in the stables now transformed into guest bedrooms. And this dead uncle was not only seen by the family but also by the neighbours regularly walking up the path from the front gate to the house.

Q. *"So you do believe in ghosts?"*

A. Shrug. Silence.

REFERENCES

Cranny-Francis, Anne (1995) *The Body in the Text* Melbourne: Melbourne University Press.

de Lauretis, Teresa (1987) *Technologies of Gender: Essays on Theory, Film and Fiction* Bloomington: Indiana University Press.

Hall, Stuart (1992) "The Question of Cultural Identity" in Stuart Hall, David Held and Tony McGrew (eds.) *Modernity and its Futures* Cambridge: Polity Press 273–325.

Haraway, Donna (1997) *Modest_Witness@Second_Millenium.FemaleMan©_Meets_OncoMouse™* New York and London: Routledge.

Heidegger, Martin (1977) *The Question Concerning Technology and Other Essays* (transl. William Lovitt) New York: Harper & Row.

Ihde, Don (1993) *Philosophy of Technology: An Introduction* New York: Paragon House.

Kuhn, Thomas (1962) *The Structure of Scientific Revolutions* Chicago: University of Chicago Press.

Latour, Bruno (1993) *We Have Never Been Modern* (transl. Catherine Porter) Cambridge, Mass.: Harvard University Press.

Michie, Elsie (1992) "From Simianized Irish to Oriental Despots: Heathcliff, Rochester and Racial Difference" in *Novel: A Forum on Fiction* 25/2: 125–40.

Rasekoala, Elizabeth (2002) "Opinion Interview" in *New Scientist* 173/2333 (9 March 2002) 44–7(1).

"Shamefully white: It's time for science to face up to an embarrassing problem" Editorial. *New Scientist* 173/2333 (9 March 2002) 3(1).

Shapin, Steven (1994) *The Social History of Truth: Civility and Science in Seventeenth-Century England* Chicago: Uni of Chicago Press.

White, Michael (2002) *Rivals: Conflict as the Fuel of Science* London: Vintage.

[JOHN POTTS]

The idea of the ghost

This essay is about the ghost as an idea that does cultural work in most, if not all, cultures. Its basic premise—that of an immaterial presence haunting the living—is the same everywhere, but it has many variants. Its precise expression is culturally determined, as the uses to which the ghost-idea is put derive from the needs of specific cultures at specific times. In some cultures the ghost may operate as a moral warning, as a signifier of respect for ancestors, as a visitor from "the other side", or as a marker of dangerous territory. In others, the ghost-idea may be taken up and used as metaphor, at the service of various discourses and their demands. In other—specifically post-Enlightenment "disenchanted"—cultures, the ghost-idea may testify to the inexplicable, in environments where mystery and magic have allegedly been explained away.

In focusing on the ghost as an idea, I am not interested in whether ghosts do or do not exist. For the purposes of cultural analysis, it is enough that many people believe in ghosts, and that even those who disbelieve know full well what a ghost is supposed to be. Everyone knows ghost characteristics. Even those who ridicule the notion are conversant with the rules of haunting, as if it were a genre. We are familiar with ghost behaviour from ghost stories, whether purely fictional or purportedly real-life experience (indeed, the slippage between fiction and alleged fact is a defining characteristic of the discourse on ghosts). We know the horror films, the urban myths, the local histories of haunted houses, the ghost stories told as teenage rites of passage. We have absorbed countless narratives of tragic ghosts, malevolent ghosts, vengeful ghosts, unhappy ghosts, angry ghosts. We know the ghost narratives so well that the general idea of the ghost is familiar to us all.

Yet the specific demands of cultural context—the various times and places in which the ghost-idea is employed—inflect the general idea in distinctive ways. This essay focuses on one of those inflections: the contemporary Western popular discourse on ghosts, flourishing most evidently on the internet. This discourse involves classifications, technological apparatus and attempts at rational analysis of phenomena. In short, it is a pseudo-scientific rendering of the ghost-idea that both preserves its general feature (an immaterial presence haunting the living) and adds particular characteristics that reflect contemporary cultural concerns. The study of this most recent inflection of the ghost-idea illuminates my general concern with the ghost as a dynamic and flexible cultural notion. Before discussing the idea of the ghost, however, I need to outline what is meant by "idea".

IDEAS IN TIME

My approach to the ghost-idea is shaped by recent developments in intellectual history emphasising the role of historical and cultural context. Earlier ventures in the history of ideas—exemplified by the work of A. O. Lovejoy (1936 and 1948)—had assumed the universality of "unit-ideas", which could be traced in a biographical fashion on their continuous path through history and across cultures. This conception of intellectual history extracted the essence of individual ideas from their march through history; these ideas were considered to be fundamentally unchanging. A succession of critiques of this formulation emerged in the second half of the twentieth century, most notably in the work of R. G. Collingwood, Thomas Kuhn, Michel Foucault, and Louis Mink. The emphasis in intellectual history since the 1960s has been on historical discontinuity and cultural difference. Histories of individual ideas have focused on the idea as an entity subject to change, dependent on its historical circumstances.

Conal Condren has summarised the influence of this impetus on the study of ideas. Drawing on Collingwood's philosophy of history, Condren defines an idea as "a compound of question and answer" (1985: 109). That is, ideas are solutions to problems, answers to specific questions asked at specific times. In this formulation of ideas, "only adequate contextualisation…can elucidate them" (1997: 51). Whereas Lovejoy and other early historians of ideas assumed the continuity of ideas as "replicated verbal formula[s]" (Condren 1985: 109), the intellectual historian attuned to historical difference regards ideas as dynamic compounds. The "verbal formula"—or the name attached to an idea—may remain unchanged, but "if the problem to which it is addressed is altered, then so is the idea" (1985: 111). Because the questions asked in one historical period differ from those posed in others, theorists such as Mink conceptualise the idea as a "time-located stage in a developmental process" (1987: 212). Cultural difference provokes further diversity in the expression of specific ideas: each culture will generate questions reflecting its own particular characteristics.

The conceptualisation of ideas as question/answer compounds helps explain the origin of individual ideas (they were answers to pressing problems), and the resistance to some ideas—even their repression—in certain historical and cultural environments. As a brief example, the history of the idea of zero is indicative of the importance of cultural context. This idea emerged in ancient Sumerian culture, becoming part of the Babylonian number system. The Ancient Greeks became aware of it by the fourth century BC, but as Charles Seife remarks, it was repressed because of its clash with "central tenets" of Western thought, primarily the cosmology of Aristotle (Seife 2000: 25). Whereas Indian and Islamic mathematics flourished, Western mathematics and science were deprived of zero until the fourteenth century, when it was finally appropriated—with much resistance from the Church—from Arabic notation. Zero, then, was initially an answer to a basic problem, that of counting. Yet its acceptance into Western culture was blocked for many centuries by an intellectual context specific to the West.

One final example illustrates discontinuity and transformation within the history of an idea. The idea of charisma was first expressed in the epistles of Paul in the first century. Emerging from the religious context of early Christian thought, "charisma" meant "gift of God's grace"—specifically spiritual gifts dispersed throughout the Christian community. This idea was repressed by the Church from the fifth century, when the mystical legacy of Pauline thought was usurped by the administrative authority of the Church. The idea of charisma lay dormant until the early twentieth century, when it was re-invented in the sociology of Max Weber. Writing in the wake of the "iron cage" of rationalisation and the "disenchanted" culture of modernity, Weber used "charisma" to mean the "certain quality of an individual personality by virtue of which he is considered extraordinary" (1968: 241). The sense of "charisma" active today is largely Weberian; applied to anyone from rock stars to politicians, it is a transformation of the idea's original—Christian—meaning.

My analysis of the ghost-idea in this essay proceeds along the theoretical lines outlined above. The ghost is an idea shaped by its worldly environment, undergoing transformations in specific cultural habitats. As it will be used to meet varying cultural demands, its functions and characteristics will alter across cultures.

The hungry ghosts of Chinese culture, for example, represent an inflection of the ghost-idea not found in contemporary Western cultures. Hungry ghosts are the unhappy ghosts of the deceased who must be ritually placated by their living family members. The hungry ghost has become malevolent because it has no direct descendants to make sacrifices to it (it may be the ghost of a child or an unmarried person). Persistent illnesses in the living family are taken to be signs of the ghost's displeasure, which must be appeased by the family securing "a descent line for the ghost" (Singer and Singer 1995: 176). This is achieved by various mechanisms, such as "postmortal" adoption, or ghost marriage, so that the unhappy ghost is provided with descendants who will recognise it as an ancestor in sacrificial rituals.

The phenomenon of the hungry ghost attests to the importance of extended family in Chinese culture. These ghosts manifest the respect for ancestors, and the responsibilities of family members to uphold tradition and the collectivity of family, even extending to the needs of the deceased. This belief performs a socially binding role in its particular culture, ensuring respect for ancestors and a sense of obligation to the past.

Earlier periods of Western culture attended to similar beliefs, as R. C. Finucane demonstrates in his cultural history of ghosts in the West. The emphasis in Ancient Greece on proper burial rites was dramatised by unhappy ghosts demanding suitable burial (Finucane 1982: 18), while medieval Europe stressed a "broad spectrum of ethical and social desiderata such as the sanctity of the grave" (1982: 86–7). By the eighteenth century, however, the "gradual loss of identification with community and extended family" was reflected in ghost behaviour: "the dead progressively withdrew from direct involvement in familial and social affairs"

(1982: 222). Contemporary Western culture, with its nuclear family base, much weaker sense of family obligation, and diminished respect for tradition, has no corresponding ghost-formation to the Chinese hungry ghost.

GHOSTS OF THE PAST

Part of my intention in this essay is to suggest the varying uses to which the ghost-idea has been put, across history and across cultures. However, the extremely widespread usage of this idea testifies to a commonality beneath the cultural variations. Finucane identifies a "treasure-trove" of ghost archetypes in Ancient Greco-Roman culture, of ghosts "informing, consoling, admonishing, and pursuing, the living"; these ghost-functions recur "down to modern times" (1982: 25–6). Unlike other ideas, such as zero or charisma, which have been specific to certain cultures and historical periods, the ghost-idea appears to be universal. It is likely that all human cultures have practised some form of belief in ghosts. Contemporary Western culture has inherited its conception of the ghost from old European predecessors (the word "ghost" derives via Old English from West Germanic); all other known cultures have corresponding words readily translated as "ghost".

An ambiguity often exists between phenomena identified as ghosts and phenomena identified as spirits or souls, an ambiguity which religious leaders have been anxious to dispel. Augustine, when defining Christian theology within a philosophical framework in the fifth century, attempted to banish pagan belief in ghosts: saints and angels may appear to the living, but the "Christian theory of ghosts" rejected apparitions of the "vulgar dead" (Tuczay 2004: 108). Despite Augustine's admonition, however, belief in ghosts other than the Holy Ghost, and reports of apparitions other than the "elite" of saints, persisted; indeed the later medieval notion of purgatory encouraged belief in traffic between the living and the dead (Finucane 1982: 58).

Despite the confusions and ambiguities occasioned by Christian theology, ghosts are commonly distinguished from souls and spirits by their manifestation to the living. That is, they are detected as apparitions, registered by the human senses (most often by sight, but also by hearing and smell). Most ancient and tribal cultures believed that the spirits of the unburied or improperly buried dead became ghosts, in the likeness of their former selves, doomed to haunt the living. It is this notion of the ghost—as an apparition representing a deceased individual or creature—that recurs across cultures. As Ludwig Wittgenstein observed, we are able to recognise this notion in cultures radically different to our own: the very existence of a word like "ghost" conveys "something in us too that speaks in support of…[tribal or "primitive"] observances." (1993: 8e) We, like Wittgenstein or even the condescending J. G. Frazer in *The Golden Bough*, understand perfectly well what earlier cultures have meant by the term "ghost" or its equivalent. While the "observances" themselves change across cultures, the idea that provoked them remains in place.

This core of the idea of the ghost, as it occurs in countless cultural contexts, is the relation to past events. With the minor exceptions of ghosts of the present (known in contemporary paranormal circles as "crisis apparitions") and the future ("harbingers"), ghosts are generally representations of the past, most commonly in the form of deceased individuals. The past may also endure in non-human form, such as animals, objects or machines. Ghost-ships and phantom objects have been reported such as the phantom railway engines that have been seen travelling along their previous routes, even once the track has been removed. Yet hauntings by deceased individuals are the most well-known type of ghost activity in which the presence of a former resident is thought to linger in a dwelling long after death.

The ghost, then, is a representation of the past as it endures in the present. To be haunted by a ghost is to be haunted by the past. Even if the ghost only exists in popular imagination, in folklore, it is keeping alive a memory of the past. In the case of haunted houses or other buildings, the ghost operates as a kind of local history. It is site-specific popular memory, a way for a community to preserve the knowledge of those who once lived there, or of that which once happened there. Ghosts can, in some cases, manifest the weight of the past, disturbing the complacency of the present in the unsettling manner of a dark or guilty secret— a shameful past. In this way, prisons, fortresses or castles are often considered haunted, preserving the memory of those who suffered and died unjustly there.

A narrative of justice frequently attends those ghosts that come back to avenge the wrongs perpetrated on them—including murder. Fisher's ghost (a widely known Australian ghost story in the nineteenth century) is of this type: the apparition of the deceased Fisher appeared, dressed as when alive, pointing to the spot where, it was then discovered, Fisher lay buried. Once the murderer was brought to justice, the ghost disappeared forever. Ghost stories such as this draw on the popular belief that official justice is often imperfect, the guilty escaping due punishment, and that sometimes action beyond the law is needed to ensure justice. In these narratives that action comes from a supernatural source, representing the power of the past to rectify the present.

Ghosts in such narratives also externalise the guilt of the living: the ghost is the burden of past actions impeding the enjoyment of the present. Ken Gelder, commenting on Australian ghost stories, refers to the prevalence, even in the nineteenth century, of narratives of displacement and dispossession: these ghost narratives draw on the way in which "white settlement is shown to be, in fact, fundamentally unsettled" (1994: xi).

If we consider the ghost-idea in the terms of Condrel and Collingwood's definition of an idea (as a question/answer compound), we must ask: what is the question to which the ghost is the answer? One possible response may be: what happens to the souls or spirits of the dead, especially those who have died in tragic circumstances? This provides only a limited definition, however, as it excludes many inflections of the ghost, such as ghosts understood to be benign. A more general, and more satisfactory, response is: how is the past kept alive? This formulation has the advantage of applying to a wide range of reported ghost phenomena. Haunted houses keep the memory of former occupants present in the minds of the living. The Chinese hungry ghosts preserve in families the memory of unfortunate ancestors. Territorial ghosts who haunt the site of their fatal accidents (also common in Chinese culture) maintain the memory of those sites as dangerous, serving as a warning to the living. Ghosts of the murdered or wrongfully executed externalise the guilt of the living. Ghosts of the unburied or of suicides preserve the memory of unhappy lives. Ghosts summoned by mediums, or who appear to loved ones, are benevolent memories of the deceased. All these differing inflections of the ghost-idea are, in their own ways, representations of the past. They keep a memory of the past alive in the present.

GHOSTS ON-LINE

I now wish to turn to the contemporary Western version of the ghost-idea, as it is enunciated, in great detail, on the WorldWideWeb. After a brief survey of the popular discourse on ghosts to be found on the Web, I will consider the cultural significance of this most recent construction of the ghost-idea.

As a public organ of popular expression, the Web hosts an extraordinary range of discourses on any topic, not least the topic of ghosts. Any search will reveal a huge volume of sites devoted to ghost research, ghost-hunting, ghost investigation, theorising about ghosts, and analysis of purported evidence—including photographs and personal accounts—about the existence of ghosts.

The status of sites varies widely, as does the mode of analysis. At the most sober end of the spectrum are sites representing academic parapsychological research initiatives. Often based at universities or other professional research organisations, these parapsychological institutions operate with strict adherence to scientific method, eschewing unfounded speculation. They are a tiny minority on the Web, a cautious voice of science overwhelmed by voices from the other end of the spectrum. At that end are found the "ghost-hunters" and the merchandisers, eager to sell ghost-hunting equipment to amateur enthusiasts. These commercial ventures sell—at high prices—"ghost detectors", ghost-hunting kits and tools, instructional courses and tour packages of haunted sites; they exploit belief in ghosts for profit.

The vast majority of sites gravitate near this sensationalist "ghost-hunting" end of the spectrum, without partaking of the cynicism exhibited by the merchandisers. Most profess a firm belief that ghosts exist. "The truth about ghosts" is offered by one such site: "Do ghosts exist? Of course they do." The more relevant question for this and many other sites is: "…granted that ghosts exist—what are they?"[1] Given the predisposed conviction, proof is not required in cases like this; however, evidence of ghostly phenomena—in the form of photographs and recordings—is in massive supply across the Web. There are many thousands of photographs posted on sites, purporting to show ghosts or other paranormal phenomena. There are also sound recordings of hauntings and Electronic Voice Phenomena (the sound of ghosts as recorded on tape) presented on sites for downloading.

The phenomena allegedly represented on the photographs are identified as ghosts if they depict human form. Other ghost-phenomena, however, are not anthropomorphic. They are assigned names such as orbs, globules, vortexes and energy anomalies, the contemporary equivalents of nineteenth-century "ectoplasm" (although that spiritualist term also survives into current usage). Orbs and other "energy anomalies" are a recent addition to ghost terminology: the International Ghost Hunters Society claims to have coined the term "orb" in 1996[2] to describe unidentified spherical figures captured in photographs.

Amateur enthusiasts have embraced these and other alleged manifestations in their excited hunt for ghosts: it is speculated that "orbs are the energy behind hauntings and at times seem to show some sort of personality."[3]

However, less sensationalist commentators have rejected these "anomalies" as nothing more than the product of low-resolution digital cameras. The more critically minded amateurs point sceptically to the link between technologies of representation and ghostly phenomena. Whereas early photography produced ghostly shapes in human form in the nineteenth century, twenty-first century digital photography has generated non-human ghostly shapes,[4] expanding the ghost-hunters' lexicon in the process. Even within the ranks of the amateur paranormal enthusiasts, the uncritical use of new technologies by a new breed of "high-tech ghost-hunters" has incurred biting criticism. One site laments the "instant gratification" yielded by photos of alleged "orbs" or "ectoplasm", often caused by elementary photographic errors.[5]

THE GHOST DATABASE

As the disputed status of "orbs" indicates, there is a diversity of approaches within amateur ghost research circles. While many sites claim photographs as conclusive endorsement of their belief in ghosts, other interpreters are a little more circumspect, judging some photos more authentic than others. Some withhold their judgement, demanding more rigorous proof. Others adopt a sceptical position, with the hope of being convinced. A frequently articulated position is that of the enlightened believer: that is, one who firmly suspects that ghosts and other supernatural phenomena do exist, but that plausible explanations must be found for them, based on sound empirical evidence.

The criss-crossing of these diverse modes makes for a curious discourse: an amalgam of the pseudo-scientific, the mystical, the sensationalist, the hopeful, the fraudulent and the fanciful. Its blend of mysticism and rationalism, like that of its more elevated nineteenth-century predecessors spiritualism and theosophy, concocts some interesting methodology. One instance of this is the frequent attempt to classify ghosts, resulting in a typology which includes the following: personal ghosts, poltergeists, crisis apparitions, harbingers, non-interactive recording ghosts, cryptic ghosts, phantoms and hauntings, as well as the ill-defined phenomena mentioned earlier, such as globules, orbs, vortexes, and energy fields.

The amateur paranormal researcher intent on classifying ghosts and related phenomena adopts the pose of a scientific researcher. The methodological endeavour displayed on such sites is intended to establish a dependable typology, to assist in the proper and responsible analysis of supernatural phenomena. The authors of these sites usually criticise the imprecise, sensationalist nature of the popular discourse on ghosts; for example, one complains of the "jumbled mess" of popular ghost terminology.[6] This site then offers a classification system that is "restricted to observable behaviours and eliminates conjecture on the part of the researcher". The aim is to produce a "database" displaying "clear, precise and retrievable justifications", enabling other researchers to focus on "what can be determined empirically".

This particular typology constructs a system of types and sub-types: "manifestations" are types of "phenomena"; a "phantom" is a sub-type of manifestation; phantoms are classified as "animate" or "inanimate"; an animate phantom may be "interactive" or "non-interactive"; they may be regular in shape or "irregular" anomalies such as orbs or vortexes. Another branch of the classification system is defined as "sentient manifestations" which are conscious, interactive presences, divided into "common" or "uncommon" sub-types; finally, "cryptic ghosts" and "ghosts", sub-categories of the common haunting, represent the "ideal manifestation" for ghost research—the ghost is "sentient, reveals itself as organic (preferably human) and its identity can be verified".

This site, like many others, imitates the rhetoric of scientific method while composing spurious classification systems. It is a prime example of pseudo-science, defined by Michael Shermer (1997: 33) as "claims presented so that they appear scientific even though they lack supporting evidence and plausibility." Shermer identifies key characteristics of pseudo-science: the use of "scientistic language and jargon" without the "evidence, experimental testing, and corroboration" required of scientific method; the dependence on anecdote rather than verifiable evidence; and the recourse to bold claims rather than scientific caution (1997: 48–49). These characteristics are abundantly displayed on ghost sites across the Web.

The alleged empirical "data" of the site mentioned above, on which the "classification of manifestations" is based, comprise "distinguishable traits that have been noted by informants and comprise documented investigations into the paranormal." That is, far from the objectively recorded empirical data demanded of scientific method, this system is founded on anecdotal accounts and mere reports of paranormal phenomena. In addition, because the typologies themselves are based on the whims of individual amateur ghost researchers, they vary substantially from site to site. Another site[7] proposes a complex system of classifications of the paranormal, involving four broad categories—"Interactive Former Human Spirit"; "Location Based Haunting"; "Psycho-kinetic"; and "Non-human Entity". Ghosts are divided into seven further classifications, ranging from Class One—an "undeveloped form" of spectre—to Class Seven—a "metaspectre with extra-dimensional powers far beyond the human ken".

Typologies such as these are often provided on expansive sites operated by amateur ghost research societies, as classification systems with which to sort "data" acquired by ghost-hunting technology. These amateur sites combine their elaborate pseudo-scientific typologies with a fascination for the technology deployed in the activities of ghost-hunting. That is, in further imitation of scientific procedure, there is a technological apparatus used to gather data. This apparatus includes: electromagnetic field detector, thermal scanner, night vision video camera, motion detector, air ion counter, audio recorder and camera. Some amateurs seem intoxicated by the technology itself: "When it comes to equipment, the more the merrier" is the claim of one ghost hunting group.[8] The more sensationalist sites accept uncritically the "data" collected by these instruments as proof of the existence of ghosts; those exercising some degree of reserve present this data as "very interesting evidence collected that lends support to the theory" that ghosts exist.[9] Needless to say, the rigorous application of scientific method—including controlled experiment and the reproducibility of experimental results—is in very short supply on these sites.

THE MODERN GHOST

This brief survey of the popular Web discourse on ghosts provides an indication of how the ghost-idea is inflected in contemporary Western culture. It is apparent that in many instances, belief in the supernatural is fused with a rhetoric drawn—very loosely—from the discourse of science. Amateur enthusiasts who firmly believe in ghosts frequently have recourse to a reasoned mode of analysis. Technological apparatus is deployed to gather data; typologies are used to assemble this "evidence". Unlike other cultures which readily accept belief in ghosts—as either mystical phenomena or culturally useful folklore—these contemporary sites need to analyse and codify the spirit world. While ancient mystical beliefs have survived, they are now conceptualised in Enlightenment terms. Belief in supernatural phenomena is filtered through a pseudo-scientific explanatory framework that makes at least some gesture to rationality.

This process generates some intriguing categories of ghosts such as the sub-type that ghost researchers have labelled "recording ghosts" or "non-interactive" ghosts. These are ghosts that are seen repetitively performing an act from their previous mortal existence, such as walking a path or climbing a staircase. They are also known as "non-interactive" ghosts because they seem oblivious to observers; unlike poltergeists, they seek no contact with the living, and seem devoid of intention or will. They are "non-sentient", lacking the power to alter their behaviour or affect their environment. They simply act out their previous behaviour as if it were a replay of a recorded event. Accordingly, they have been likened, in the popular literature of ghost research, to tape recordings, cinema replays, memory traces, astral recordings and psychic imprints.[10]

In one sense, this concept partakes of the ancient mystical traditions of spirit of place and earth memory. The Celts believed that all experience is "remembered" by the earth, and that trees especially, were repositories of memories and spirits. Other cultures believed that ghosts were lodged in certain substances such as particular stones. Recording ghosts, in a similar fashion, are thought to exist as a form of recording—or memory—imprinted in the stone and other materials of buildings.

Yet there is something specifically contemporary about the version of mystical tradition found in the theory of "recording ghosts". Its quasi-scientific approach marks it as a post-Enlightenment speculation, while the centrality of recording as metaphor places it firmly within the late twentieth and early twenty-first centuries. Historically, ghosts have been conceptualised according to the prevailing cultural currents of the time; in the contemporary environment, that includes the significance of recording and information—as technology and metaphor. Hence we have "non-interactive" recording ghosts which play and replay like computer programs or video recordings.

The contemporary ghost-formation takes its place within (at least) two continuities: the endurance, from ancient times, of the ghost-idea itself; and the strain of media mysticism stretching back to the early nineteenth century, as documented by Jeffrey Sconce in *Haunted Media* (2000). Photography, cinema and radio were all enshrouded in a mystique; ghosts inhabited all these devices, and were recorded by them. It should surprise no-one that computer networks—founded on the immateriality of digital information—are haunted by the sights and sounds of ghosts. The Web is also the home of the modern ghost-hunters, convinced of the reality of their prey, determined to capture the immaterial with apparatus and recording technology, and driven to explain the phenomena they encounter in a discourse crudely approximating that of science.

The contemporary ghost—as charted, recorded, depicted, analysed, classified and explained on the Web—performs complex cultural work. It attests to the persistence of mystical belief in societies founded on rationalist principles. It demonstrates that modernity was never fully "disenchanted". It reveals that enchantment—in the form of belief in the supernatural—flourishes even in highly technologised cultures. And the determination of the ghost hunters and classifiers to codify their prey in pseudo-scientific terms shows how the ghost-idea has been adapted to suit the demands of the present. Contemporary ghosts are constructed as both supernatural and scientific phenomena. They take on forms—"orbs" and "vortexes"—generated by the very technologies used to reveal them. They perform the work of fusing magic and technology, of bridging mysticism and science. They are hybrid ghosts, in an age that wants to believe in the powers of science but is drawn with an equal force by the lure of the unknown.

ENDNOTES

1. "The Truth About Ghosts" by Donald Tyson at http://www.ghosts-uk.net/modules/news/article.php?storyid=72 accessed 14 April 2004.
2. http://www.ghostweb.com/orb_theory.html accessed 25 May 2003.
3. The Amateur Ghost Hunters of Seattle http://axgxhxoxsxtx.tripod.com/hottopics/idl.html accessed 22 May 2003.
4. Bryan Bonner "Spirit Photography: Digital vs Traditional & The Mystery of Orbs" at http://rockymountainparanormal.com/orbarticle.htm accessed 24 May 2003.
5. "Ectoplasm: Real or Hoax" http://www.rockymountainsparanormal.com/Ectoplasm.html accessed 28 May 2003.
6. "Classification of manifestations" by Christopher J. Williams at http://www.rockymountainparanormal.com/classification.htm accessed 14 April 2004.
7. The Paranormal and Ghost Society at http://www.paranormalghostsociety.org/ghosts.htm accessed 14 April 2004.
8. The Philadelphia Ghost Hunters Alliance at http://www.members.aol.com/Rayd8em/Eqipment.html accessed 20 May 2003.

9. Ghosthounds: Paranormal Investigations—Paranormal Productions at http://www. ghosthonds.com accessed 20 May 2003.
10. Anthony North (1998: 22) discusses the "tape-recording" theory of ghosts, as do Hilary Evans and Patrick Huyghe (2000: 141).

REFERENCES

Collingwood, R. G. (1946) *The Idea of History* Oxford: Oxford University Press.

Condren, Conal (1985) *The Status and Appraisal of Classic Texts: An Essay on Political Theory, Its Inheritance, and The History of Ideas* Princeton: Princeton University Press.

---------------(1997) "Political Theory and the Problems of Anachronism" in Andrew Vincent (ed.) *Political Theory: Tradition and Diversity* Cambridge: Cambridge University Press.

Evans, Hilary and Huyghe, Patrick (2000) *The Field Guide to Ghosts and Other Apparitions* New York: Quill.

Finucane, R.C. (1982) *Appearances of the Dead: A Cultural History of Ghosts* London: Junction Books.

Gelder, Ken (ed.) (1994) "Introduction" in *The Oxford Book of Australian Ghost Stories* Melbourne: Oxford University Press.

Lovejoy, Arthur O. (1936) *The Great Chain of Being: A Study of the History of an Idea* Cambridge, Mass.: Harvard University Press.

-----------------(1948) *Essays in the History of Ideas* Baltimore, Maryland: The Johns Hopkins Press.

Mink, Louis O. (1987) *Historical Understanding* B. Fay, E. O. Golob and R. T. Vann (eds.) Ithaca: Cornell University Press.

North, Andrew (1998) *The Supernatural: A Guide to Mysticism and the Occult* London: Blandford.

Sconce, Jeffrey (2000) *Haunted Media: Electronic Presence from Telegraphy to Television* Durham: Duke University Press.

Seife, Charles (2000) *Zero: The Biography of a Dangerous Idea* London: Souvenir Press.

Shermer, Michael (1997) *Why People Believe Weird Things: Pseudoscience, Superstition, and Other Confusions of Our Time* New York: W. H. Freeman and Company.

Singer, André and Singer, Lynette (1995) *Divine Magic: The World of the Supernatural* London: Boxtree.

Tuczay, Christa A. (2004) "Interactions with Apparitions, Ghosts, and Revenants in Ancient and Medieval Sources" in James Houran (ed.) *From Shaman to Scientist: Essays on Humanity's Search for Spirits* Lanham: Scarecrow Press.

Weber, Max (1968) *Economy and Society: An Outline of Interpretive Sociology* G. Roth and C. Wittich (eds.) New York: Bedminster Press.

Wittgenstein, Ludwig (1993) *Remarks on Frazer's* Golden Bough (transl. A. C. Miles) Denton: The Brynmill Press.

[STEPHEN MUECKE]

Contingency and ritual on the island of ghosts: new ethnography in Madagascar?

"SHE DANCED AWAY ON THE OTHER SIDE"

When William Ellis, missionary and photographer, was sent to Madagascar in the mid-nineteenth century by the London Missionary Society (LMS), he was serving both God and Science. In these early days of photography, the technology was spreading rapidly throughout the European empires, and images such as Ellis', along with all the cargo of scientific artefacts and writings, had the spectacular effect, in Europe, of reporting back on other peoples encountered.

But Ellis's greatest challenges were in the field where his adventitious work was assailed by political contingency. The protestant LMS was in competition with the French Jesuits for access to Madagascan souls. Earlier, in 1835, Queen Ranavalona I had been executing Christians en masse, and had closed down access to foreigners but by 1852 news came through that the country was opening up again under her son, the successor to the throne, Prince Rakotondradama. The race was on to influence the prince, and the strategies were to be overtly technological but, for safety's sake, only covertly Christian: "The more recent and spectacular inventions of Western technology were regarded as particularly useful means of gaining access by baffling and impressing" (Peers 1997: 25). Ellis had prepared himself by learning photography but had to wait for three years at Mauritius for his chance to reach the centre of power in Madagascar. Meanwhile his Jesuit rival Père Marc Finaz had gained access to the Madagascan capital under a false identity, and set about pre-empting, with his own machines, what he knew would be Ellis' technological assault; he "adopted the role of concert pianist and experimented with hot air balloons" (Peers 1997: 25). Knowing that Ellis had a camera, he sent for a daguerreotype himself and as soon as it arrived he worked day and night "to gratify those people who will be useful to me" (1997: 25). But he did not know that Ellis had already been sending albumenised salt prints from Mauritius to the Madagascan capital, and these certainly facilitated Ellis' eventual access. The camera was indeed, as he said, "the apparatus necessary for working miracles according to the improvements of modern science" (1997: 25).

But this Western magic encountered a Madagascan counter-magic. When the weak young prince began to lose power the tide turned once again against foreigners, their religions and their other trappings. There was a strange manifestation of choreomania, dancing of the possessed, or "ghost dancing", which the foreigners struggled to understand. But Ellis knew, at any rate, that it interfered with his photography:

> On two occasions I was somewhat troubled with them [dancers], first when I was taking a photograph of the site of the martyr's death at Ambohipotsy. I had fixed my camera and was engaged in the tent, when my servant who I had left to watch called out that a sick dancer was coming. She was a decently dressed young woman...I stood by my tent and told my servant to hold the camera stand. She danced once or twice around the camera, coming nearer each time and she put out her hand as if to take hold of it, I hastened towards it when she danced away on the other side and went dancing down the hill (Peers 1997: 28).

This story illustrates for me what it is about the concept of contingency that suggests new ethnographic practice. If earlier ethnographers were equipped with God and Science—pillars of the "modernist settlement" of European rationality (Latour 1997: xi–xii)—it allowed them to *circumscribe* and *exclude, in the declaration that this* will be my field, my community, or my tribe. I draw a line around it. *These* other questions will not be relevant. Systematic purification of the field of data will decide what is my *necessity* (taking a photograph, doing a study of kinship), and exclude the *contingent* (that scary dancing woman).

By contrast, treating the world as a complex open system, with new ethnographic techniques, means one is alert to the *feeling* that there might be something *else* there that could transform the research agenda. Contingency, in its Latin root, is about touching, bordering on, reaching, befalling. It is not therefore about maintaining critical distance, but about tipping over into new paradigms where encounters (with Others, for instance) can teach us, not necessarily by direct instruction, but by putting our preconceived ideas in jeopardy. It is about not eliminating the risky or the accidental.

What do I mean by contingency? The word has a Nietzschean connection, to embrace contingency is to recognise multiple and hidden causes. It is about "a quarrel between poetry and philosophy, the tension between an effort to achieve self-creation by the recognition of contingency and an effort to achieve universality by the transcendence of contingency" (Rorty 1998: 25). This transcendental philosophical effort "to see life steadily and to see it whole" (1998: 26) is an immense fiction, a striving for unity. Similarly, when it applies in ethnographic practice to ritual it is about the elimination of all contingencies so that the ritual can focus on its necessary outcome. Put another way, in everyday life someone may wander into the kitchen thinking she or he has the intention to get a glass of beer, but she or he ends up choosing a coffee. Everyday life is thus full of contingencies which do not seem to matter. But in a ritual or ceremony, like a wedding, it is crucial for the participants that all contingencies be removed, or planned for, so that the desired outcome (what is necessary for the event) be achieved. But I find, for a new ethnography, that I am prioritising the contingent. To the extent that it becomes the *necessary* concept, I arrive at the paradoxical: the erstwhile necessary will turn out to be dispensable, and the contingent will become necessary in the sense that it is the place from which all future potential is broadcast.

The contingent is thus about potential as it links things unexpectedly together. It does not seek to continue the positivism of an anthropological practice which constructs another society as unified, "over there" objective, and characterised by "certain distinctive beliefs". The new ethnographic method will work by way of connection and articulation. There will, in this method, be no way in which "they" remain superstitious, about, say, an eclipse of the sun, while "I" am necessarily beyond that historical stage. That would be to go along with another colonialist story of historical seriality where civilisations are lined up on a scale of technological progress.

My trip to Madagascar takes shape—in writing—during and after the actual voyage, the trip is a pretext, naturally, for a story and an argument. The argument, about contingency, starts like this: the subject (you and I) is formed out of contingent elements—your parents, the circumstances of your life, the choices you make, and so on.[1] One anthropologist discovers just this kind of fluid identity among fishing people on the west coast of Madagascar, using my word for the day: "The Vezo's 'identity' does not become a permanent feature of the person but remains contingent on what the person does in a certain place and at a certain point in time" (Astuti 1995: 472) On the so-called objective side, what makes history is the unpredictable surprise of the accident, significant enough to become an event. Only the planned, predictable and repetitive are forgotten. So in the end what we call our societies and cultures are a haphazard combination of these two levels of contingency: that of the self and that of history. Haphazard, but somehow it functions. Mostly with a weak sense of forward movement, towards death in the case of the individual, towards destiny in the case of society, both have some narrative organisation which we relate to each other in endlessly repeated stories.

EXODUS FROM LEMURIA

My story begins, by chance, in Australia, where I have recently read a statement from an Indigenous Australian talking about the origins of his people. David Unaipon was writing in the 1920s:

> Nearly all the tribes scattered about Australia have traditions of their flight from a land in the Nor-West, beyond the sea, in Australia. That land may probably be the ancient continent of Lemuria. The traditions also relate that the aboriginals were driven into Australia by a plague of fierce ants, or by a prehistoric race as fierce and as innumerable as ants. Like the Israelites aboriginals seem to have had a Moses, a law-giver, a leader, who guided them in their Exodus from Lemuria. His name is Narroondarie (Unaipon 2001: 4).

This Lemuria is a mystical land, also known as Mu, well-loved by the New Age industry but it is also the name shared by Madagascar and its proto-primates which we call lemurs. Where did they get this name?[2] Was it because their weird cries struck European explorers as the cries of the dead, and that they were recalling the ancient Roman rites as described by Ovid in his *Fasti*?

> When from that day the Evening Star shall thrice have shown his beauteous face, and thrice the vanquished stars shall have retreated before Phoebus, there will be celebrated an olden rite, the nocturnal Lemuria: it will bring offerings to the silent ghosts. The year was formerly shorter, and the pious rites of purification (februa) were unknown, and thou, two-headed Janus, was not the leader of the months. Yet even then people brought gifts to the ashes of the dead, as their due, and the grandson paid his respects to the tomb of his buried grandsire. It was the month of May, so named after our forefathers (maiores), and it still retains part of the ancient custom. When midnight has come and lends silence to sleep, and dogs and all ye fowls are hushed, the worshipper bears the olden rite in mind and fears the gods arises; no knots constrict his feet; and he makes a sign with his thumb in the middle of his closed fingers,[3] lest in his silence and unsubstantial shade should meet him. And after washing his hands clean in spring water, he turns, and first he receives black beans and throws them away with his face averted; but while he throws them he says: 'These I cast; with these beans I redeem me and mine.' This he says nine times, without looking back: the shade is thought to gather the beans, and to follow unseen behind. Again he touches water, and clashes Temesan bronze, and asks the shade to go out of his house. When he has said nine times, 'ghosts of my fathers, go forth!' he looks back, and thinks that he has duly performed the sacred rites (Ovid 1967: 291–2).

Ovid speculates that the name has its origins in the foundation of Rome, the city which took its name from the man who murdered his brother, Remus. In order to propitiate the spirit of Remus, Romulus agreed to give:

> …the name of Remuria to the day on which due worship is paid to buried ancestors. In course of ages the rough letter, which stood at the beginning of the name, was changed into the smooth; and soon the souls of the silent multitude were also called *Lemures*: that is the meaning of the word, that is the force of the expression (Ovid 1967: 295–7).

The more I want to relate to you a simple story about Madagascar, the more complex it becomes. I cannot disentangle the country from the signs which surround it, these spurious contingent meanings which connect it to Ancient

Rome, and even to the mystical land of Lemuria. Madagascar begins to disappear over the horizon of orthodox reason, to reappear as paradox. In the Western onto-theological tradition, paradox, like contradiction, occupies the place where reason cannot go. It seems even the name "Madagascar" was a malostension made by another Italian, Marco Polo, as he voyaged around the Indian Ocean, without even reaching the large island:

> One of Marco's worst errors was to mix up Madagascar and Mogadishu in the Horn of Africa: "The meat eaten here is only camel flesh. The number of camels slaughtered every day is so great that no one who has not seen it for himself could credit the report of it." This is exactly true of Mogadishu, but certainly not of the great island 2000 miles [3220 km] to the south. (It is testimony to the influence of Marco Polo that the name Madagascar, taken directly from his writings, has survived despite being based on a total confusion) (Hall 1998: 52).

Thus chance has determined the fate of every Malagasy whose name emerges out of the contingency of European exploration. But also their own, for the present inhabitants of Madagascar are the descendants of Indonesians who arrived on the island some 1500 years ago. The language is still a cousin to the Indonesian language, the people look Indonesian, and their funeral ceremonies link back to possible early Indonesian ancestor-worship.

But let me summarise so far. Madagascar was part of an ancient continent that some came to call Lemuria. Its unique primates are called lemurs, or in French, *lemuriens*. Both relate to ghosts and the dead, the lemurs were named after Lemuria which is a Roman ritual for propitiating the spirits of dead relatives. The lemurs were so named because their strange cries heard at night were thought to sound ghostly. Did those in Lemuria who named the lemurs do so in ignorance of the Malagasy custom of reburial to honour and propitiate the spirits of the ancestors? I doubt it. You can buy a CD by the famous Malagasy musician Rossy called *Island of Ghosts*. And there is a documentary film of the same name. Now, these are convergences, perhaps, rather than divergent contingencies. But they nonetheless already break out of the modes of classification which our sciences will normally permit. You cannot talk about Roman rituals, primates and contemporary World Music in the same paragraph.

Contingency, Nicholas Smith (1997) tells us, is a form of "strong hermeneutics". Now what this might mean is that the interpretative process is a creative one allowing a person to "make connections" in the processes of perception, observation, thought and writing. Interpretation through contingency re-sorts sets of data which have been sorted out into separate categories by scientific method, like science and religion, like magic and technology. It goes backwards through history to rediscover the kinds of connections made by people in their everyday life, such life which is full of inexplicable contingencies. It can of course make false connections; these will have no historical resonance and probably disappear. But it is the case that most "breakthroughs", in even the hardest sciences, are made through creative experimentation, as opposed to the repetitive "non-experimentation" which expects things to respond to our questions as reliable witnesses, and our machines to have exorcised their ghosts so that they simplify and purify data, eliminating the unreliable and the contingent. It is the unreliability of the machine which can be productive of the ghostly presences, and so too can its cultural excesses: William Ellis' photography in Madagascar, far from being a reliable witness, was a miracle-machine in a political contest for another intangible, Christian faith.

TURNING THE BONES

You can dismiss the beliefs of other peoples if you like, and this can be refreshing, as when William Burroughs says about Madagascar there are "worthless zebus, a small hump-backed breed of ox venerated by the natives and tied into some idiotic funeral practices (1995: 32). Making fun of natives is sport for the philosopher-pirates, but what we are dealing with here is the Madagascan *famadihana*, the exhumation and turning of the bones of the ancestors, a veritable destiny for the life of every important Madagascan whose ambition is to die well, and to die in the right place. To be venerated, to be buried, and to hope to be exhumed in the tomb of the village of one's fathers where the best and most important buildings, typically again, are the tombs. Here after seven years you hope you will be important enough for a *famadihana*, that there will be a contest between the quick and the dead, as the living pay a lot of homage to the ancestors, going to them in a procession full of fear and getting reckless with the rum, to a music which intensifies and eventually eroticises the relationship, the skeletons are rewrapped, placed in the laps of women whose fertility is then assured, they will dance with these cadavers, they will get careless as even more rum is consumed, the wrapped packages will fall, the bones will be crushed under dancing feet and the whole thing will culminate in a scramble for the mats on which the dead reposed; women treasure these sleeping mats for their fertile properties.

So Westerners must believe in something after all because we find we too cannot be "absolutely modern", and just dispose of the dead and forget about them. Funeral procedures are what archaeologists reckon distinguish humans from the animals from the very beginning. They locate the most primitive human societies by finding ancient bones which show signs of cremation or sacrifice. So what then is the "status" of the dead, generally or theoretically? We cannot do without them completely, otherwise we would "revert" to the status of animals (which just interact without producing that magical surplus to their activities known as culture), nor can we imagine them being with us all the time (—Where's your Mum today?—Oh, she'll be back soon). If we do, we are mad, or possessed, like the Vezo people, with a *tromba* [ghost].

Dismissing the beliefs and practices of others as "idiotic" is almost exactly the same as saying that they are "interesting", which is the usual job of liberal-minded social scientists, treasuring a little bit of difference as it is labelled and exhibited in the Museum of Lost Beliefs (as Burroughs might put it) in a great city of the global north. I would claim that the radical contingency I am advocating will only allow you to talk with the "natives" in such a way that their agenda touches yours and vice versa. But it also gets you into other sorts of trouble.

RETHINKING *LA FEMME MALGACHE*

I will not get to see a *famadihana* on my trip to Madagascar. I will not even get to see a ghost, but I have my copy of William Burroughs' *Ghost of Chance* with me, and by chance I stay at the Hotel Indri on the first night. Burroughs' Captain Mission, the pirate-philosopher, I read in bed:

> …had smoked opium and hashish and had used a drug called *yagé* by the Indians of South America. There must, he decided, be a special drug peculiar to this huge island, where there were so many creatures and plants not found anywhere else. After some inquiries he found that such a drug did exist: it was extracted from a parasitic fungus that grew only on a certain spiny plant found in the arid regions of the south.
>
> The drug was called *indri*, which meant "look there" in the native language. [My hotel had a big mural of a lemur on the side, the species called *indri*, but

Burroughs is a fiction writer, he can make things up.] For five gold florins he obtained a small supply from a friendly native. The drug was in the form of greenish-yellow crystals. The man…showed him, exactly how much to take and cautioned him against taking any more.

"Many take *indri* and see nothing different. Then they take more and see too much different."

"Is this a day drug or a night drug?"

"Best at dawn and twilight" (Burroughs 1995: 10).

I woke at dawn to the noises of the jungle, the ecstatic cries of an amorous duet in the next room. Could the *Hotel Indri* be a *hotel de passe*?

I start to learn things, I take a walk in the early morning and a young woman crouched in a doorway not far from the French Embassy shows me, with a big amused smile, a page from a porn magazine illustrating fellatio. I start to think travelling alone is not a good idea at my age. What I mean is, all the other middle-aged European-looking men in Antananarivo seem to be sex tourists looking for young girls, something the government is campaigning fruitlessly against: the low HIV rate in Madagascar attracts the men with foreign currency. There is nothing for me to do in the evenings, the most interesting place is the cabaret at the Hotel Glacier where they play local malgache tunes and popular rock covers. And where the prostitutes continue to think that my refusal to provide them with their due income can only be due to some perversity.

My day-to-day work involves making contact with other researchers for a research project I was doing on culture and commerce in the Indian Ocean. When I came to pitch this project to the *chercheurs* (and *chercheuses*) of the Institut de Civilisations, and had offered them a special issue of the *UTS Review* (on the Indian Ocean) as a gift, they had to offer me in return a copy of their journal, *Taloha* ("Times Past"), the issue on "Rethinking la femme Malgache". The director remarks in a jocular fashion as he hands me the book, in front of the whole committee: "I don't know if Monsieur has had the chance to *experimenter* any Malgache women yet"!

My only experiments have involved shrugging off the street women near my new hotel, in a classier part of town. This evening I am pursued by two, I say I have no money, then realise too late that this is some kind of admission, and their heels go click-clack-click across the cobblestones as they come after me again.

"Leave me alone", I protest, "what do you want?" "We will do *whatever* you want, Monsieur," one says with a little shimmy of her hips.

That night I dream that I am feeding my cat in a kitchen somewhere, and suddenly there are five or six black cats, coming in the window. One jumps up on my shoulder, as I lean away, holding the bag of pet food out of reach.

In the morning I pick up the copy of *Taloha* 13, "Rethinking la femme Malgache: new views on gender in Madagascar", and start reading "The Rights and Wrongs of Loin-Washing", by Karen Middleton, of Oxford. It begins:

> Among the Karembola of Southern Madagascar, a woman is entitled to demand a sacrifice to purify her body when her husband sleeps with a "stranger", a woman unrelated to the wife. This paper describes some of the cultural values that lie behind the practice of "loin-washing" (*sasa valahañe*), and seeks to understand why Karembola represent this simple rite in contradictory ways. It relates these contradictions to broader paradoxes in Karembola communities, and in so doing highlights the contribution women make to local political and social processes through their sexual politics and exchange relationships.

What does this little bit of social scientific prose do? It unifies, and separates off, the Other. As she writes, Middleton is immediately, suddenly, "among the Karembola" but there is no line of flight, how she got there, or how she keeps linking back. It is in the eternal present tense, it is not an historical process, nothing is expected to change. There are "contradictions" and "paradoxes," but I would bet the Karembola do not talk about their practices in these terms. These are the philosophical terms which Middleton wants to resolve in the vocabulary of the social sciences, using words like, "represent", "social and political processes", "cultural values" and "exchange relationships".

These are the common analytical strategies in the social sciences that seek to see through the surface of things to give us access to a level of reality somehow deeper than the everyday (Chakrabarty 2000: 239). But in the process it hides from view the fragmentary nature of the "now" that the investigating subject inhabits.

"AN INDESCRIBABLE ABSTRACTED EXPRESSION"

The Madagascans, obsessed as they are with death and their ancestors, become possessed—some of them—by spirits of the dead. They may come to them in dreams, or they may permanently occupy individuals. In the northern parts of the island, the possessed are most often older single women, and they are possessed by spirits who are called *tromba*, who are actually royal ancestors. The possessed might dress up in the style and period of the person inside them. James Sibree writing in the later nineteenth century talks of a different kind of possession, one that returns us to the story about the photographer at the beginning, and expands that clash of technological magic and corporeal magic into an issue to do more with an "ecology of practices" as Isabelle Stengers (2002: 262) calls it ("how different forms of knowledge and cultural practices work"):

In the month of February 1863, the Europeans resident at Antananarivo (Tananarive), the capital of Madagascar, began to hear rumours of a new disease, which it was said had appeared in the west or south-west. ...After a time...it reached the capital, and in the month of March began to be common. At first, parties of two or three were to be seen, accompanied by musicians and other attendants, dancing in the public places; and in a few weeks these had increased to hundreds, so that one could not go out-of-doors without meeting bands of these dances. It spread rapidly, as by a sort of infection, even to the most remote villages in the central province of Imerina...

The public mind was in a state of excitement at that time, on account of the remarkable political and social changes introduced by the late king, Radama II. A pretty strong anti-Christian, anti-European party had arisen, who were opposed to progress and change. This strange epidemic got into sympathy, especially in the capital, with this party, and the native Christians had no difficulty in recognising it as a demoniacal possession (...)

The patients usually complained of a weight or pain in the praecordia, and great uneasiness, sometimes a stiffness, about the nape of the neck. Others, in addition, had pains in the back and the limbs, and in most cases there seems to have been an excited state of the circulation, and occasionally even mild febrile symptoms (...) they became restless and nervous, and if excited in any way, more especially if they happened to hear the sound of music or singing, they got perfectly uncontrollable, and, bursting away from all restraint, escaped from their pursuers and joined the music, when they danced, sometimes for hours together, with amazing rapidity (...) The eyes were wild, and the whole countenance assumed an indescribable abstracted expression, as if their attention was completely taken off what was going on around them. The dancing was regulated very much by the music, which was always the quickest possible—it never seemed to be quick enough. It often became more of a leaping than a dancing. Thus they danced to the astonishment of all, as if possessed by some evil spirit, and with almost superhuman endurance—exhausting the patience of the musicians, who often relieved each other by turns—then fell down suddenly, as if dead; or, as often happened, if the music was interrupted, they would suddenly rush off as if seized by some new impulse, and continue running, until they fell down, almost or entirely insensible. After being completely exhausted in this way, the patients were taken home, the morbid impulse apparently, in many cases, destroyed (…) Many of them professed to have intercourse with the departed, and more particularly with the late queen (Bloch 1971: 21–4).

James Sibree, again with the LMS, had just arrived in the capital, so was an eye-witness. He writes well about this phenomenon, but has he been talking to doctors? No doubt, since the dancers are already referred to as "patients", and there are technical terms like "praecordia". The other discourses intersecting here concern religion and politics, revealing perhaps even more about the phenomenon than the medical discourses, but more importantly, these (somewhat incompatible) causes are *multiplied*, and the syntax betrays a desire to entertain the collective singular: "the public mind was in a state of excitement…", he says.

Bloch, the anthropologist who cited Sibree 100 years later, has things to add from his discipline, weaving yet more threads into the multiple causality: "What Sibree fails to make clear is that the dancers apparently believed that they were preparing for the return of the recently dead traditionalist queen, Ranavalona" (Bloch 1971: 24). As noted above, Madagascan cultures are strongly centred around cults of the dead, and the major ceremony (*famadihana*) involves ritual reburial where the power of the ancestor is made manifest. In the case of spirit possession by royalty, this dancing mania may well be thought of as a kind of collective re-embodiment. And Bloch notes that (in addition to the crucial element of music which makes the events social and planned rather than accidental), the movements are *meaningful*:

…The dancers believed they were carrying [the Queen's] baggage from the coast to the capital and they mimed carrying heavy loads which they passed one to the other in relay…it recalls the famous cargo cults and other millenarian cults which it has often been convincingly argued are closely linked to violent foreign contact. In this case, however, the somewhat surprising element is the absence of any clear leaders of any kind, and perhaps the closest parallel is therefore with the ghost dancers of North America... The significant aspect of the movement is the combination of the frenzied situation, reaffirmation of the past, the focus on tombs and the dead and the total rejection of western influences (Bloch 1971: 25).

The rejection of things Western was manifested in the dancers, or perhaps we could now call them, anachronistically, demonstrators, knocking off people's hats and killing pigs, both Western imports. The dancers were not therefore self-contained, they were interacting strongly with bystanders, forcing them to greet them, smashing contents of houses, and encouraging them to join the movement, which even a parade of soldiers did, going into a frenzy and attacking their officers (Andrianjafy 1902: 60). The anthropological comparison with the ghost dances of North America encourages us to think of this as an anti-colonial protest, but that particular global political framing was certainly not available to the locals.

The local newspaper, the *Moniteur Universel*, reported on 7 Juillet 1863 that the dancers:

said that Ranavalo et Radama 1st emerged from their tomb to declare their son unworthy of the crown. It was said he had sold his country to the whites. His mother and father were groaning under the weight of this monumental crime. Their spirits were weeping and beseeching all their old subjects to seek the help of *sikidys* (diviners) so as to deflect the curses cast by on their unfortunate successor (Andrianjafy 1902: 61).

The expatriate doctor Andrianjafy wrote a thesis about this phenomenon in Montpellier in 1902 remembering what he had observed as a boy, and gave it a medical explanation: an origin in malarial infection. But at the same time he saw that it was literally orchestrated; musicians had to be there playing fast music for the dancers to "cure" them. His prescription was not only treatment for malaria, but also as a second step, *modernisation of the whole society*, including an "uncompromising rout of witches, diviners and others exploiting the credulity of the people. Ancestral prejudices, superstition and ignorance, which are jealously maintained by these people, must be eliminated by progress, civilisation and science" (1902: 62). Towards his conclusion his rhetoric becomes more strident: "This is a battle for moral hygiene on three fronts: political, religious and social" (1902: 63).

Ian Hacking, in his book *Mad Travelers*, looks at "transient mental illness", in particular the fugue syndrome appearing in France about the time of the invention of the bicycle. The bicycle, an ideal mode for individual escape, is to the fugue as music is to choreomania: an exciting technical prosthesis. All factors must be looked at for an understanding of these phenomena such that Hacking's evocation of the "ecological niche" is useful in that it invites the kind of *complex* analysis proper to ecological studies. But he also finds major cultural tensions in operation, saying that one of the vectors within which mental illnesses find themselves is "cultural polarity: the illness should be situated between two elements of contemporary culture, one romantic and virtuous, and the other vicious and tending to crime" (Hacking 1998: 2). For the Middle Ages such a cultural and social polarity finds its sanctioned reversal in Rabelaisian "carnival", for which other dancing manias like this one—St Vitus Dance, the Tarantella— may well have been dangerous precursors.[4] In the case of Madagascar the cultural polarity is the one set up by colonialism and the late queen's rejection of it. The cult wants to bring her back from the dead to fight the battle again, and to reject the foreigners. But it is too late, Madagascar from this moment on will enter the colonial world, like that British sun that never sets.

ECLIPSE TOTALE

I ask the clerk at the Hotel Indri, "what are the *ombiasy* (healers and astrologers) saying about the eclipse?" He does not know, but refuses to take an anthropological view on the matter, that is, assessing the beliefs of his community as if Other. "It is a natural phenomenon", he says, using what he thinks is my belief system, and his, as non-superstitious modern people.

Diary entry, 21ˢᵗ June: Went down to Ave de la Liberation. Very few people about, cars off the street because the government feared "hot-heads" would get out of control (see paper, see also editorial). Quite a few "normal" sunglasses indicating a nervousness, as did one family with the protective eclipse-watching glasses taped to their two little boys' heads, so that the elder boy seemed to have to watch the sun all the time, and the little one was looking away, seeing nothing through the opaque material.

Another family went past quickly at the height of the eclipse, the woman with a shawl completely covering her face, the man holding up his newspaper to protect himself, they hurried on.

A Frenchman turned up and sat down on the steps of the liberation monument in a desultory fashion. A Malagasy asked him if he had the special glasses. *J'en ai, les lunettes,* "I have the glasses." Had he looked at the sun yet? No answer. The Frenchman sat with his back to the sun and never looked, as if he sat in silent critique of the superstitions and fears of the locals.

The BBC world service to Africa that evening reported fireworks and celebrations in Zambia, but the correspondent from the little village in the west of Madagascar said that the majority of the people had stayed inside on the advice of the *ombiasy*. A lunar eclipse, the local astrologer/healers had concurred, "could be very good or very bad".

For Richard Rorty the process of enlightenment secularity should continue so that the world would be "de-divinised" (1989: 45) so that we would be cured of any deep metaphysical needs and "recognise the contingency of the vocabulary in which [we] state our highest hopes" (1989: 46). The final result would be that we would "no longer be able to see any use for the notion that finite, mortal, contingently existing human beings might derive the meanings of their lives from anything except other finite, mortal, contingently existing human beings". Rorty's world is too congenial, everyone in it must be North American College-educated. The real people of the world are full of superstitions and religious beliefs. Gods still perhaps inhabit our cathedrals, and spirits are down the bottom of the garden. In Australia, a man of the cloth was appointed to the position of governor-general.

Writing in his new book *Provincializing Europe,* Dipesh Chakrabarty tells us "why he is not a secularist", to borrow the title of William Connolly's latest book Dipesh says that there is an:

> Assumption running through modern European political thought and the social sciences…that the human is ontologically singular, that gods and spirits are in the end "social facts," that the social somehow existed prior to them. I try, on the other hand, to think without the assumption of even a logical priority of the social. One empirically knows of no society in which humans have existed without gods and spirits accompanying them. Although the god of monotheism may have taken a few knocks—if not actually "died"—in the nineteenth-century European story of the "disenchantment of the world," the gods and other agents inhabiting practices of so-called "superstition" have never died anywhere. I take gods and spirits to be existentially coeval with the human, and think from the assumption that the question of being human involves the question of being with gods and spirits (2000: 16).

This is not asking for a return to the metaphysical, or towards anything New Age, it is rather asking us to accept that many people are motivated by whole bundles of contingent forces which are not seen as limited to the purified rational realm of the secular, mortal and finite.

What if our starting point is not community, defined by its borders and its inward self-defining gaze, but the approach of the stranger, and the connectedness he or she brings from the outside? Was it not always so for the anthropologist or writer to be able to report on the other society, that they came in on a line: an airline, a railway line, and they would never be cut off completely and thus lose their identity, there was always some line back, like a phone line. And they reported on sameness and difference. *This* is what makes Merina culture in the Madagascan high plateaus unique, *but* there are similarities with us, (whoever we are), and in this play of sameness and difference a tale will be told with the inevitable conclusion justifying the imperialism of a way of knowing about others: there is something the same deep down, our very humanity perhaps, and that the Western scholar has the right to pronounce on it.

That banal conclusion is nevertheless the foundation of a way of knowing which the anthropologists would be quick to deny and banish: rather the importance will be in the descriptive details. True, and the details are important, but when, when do they cease being contingent? Only when the subject is in the overarching position of making cultural comparisons, and equipped with some system for building up the details into a picture of a self-contained community. Rather, I want to work the connections, nothing but the lines. The international flights, the lines of communication (phone, radio, television, internet, and the like, print publications), shipping and trade, the movement of people; all this both inside and outside of the borders of a given community or nation. This is an approach which is more in tune with globalisation and its resistances. It takes the relationship as primary and the entity as secondary. We can examine these relationships at their points of intersection which are what I will call points of contingency.

I have argued that the grasp of the radically contingent is the precondition for developing a newer vocabulary in our academic writing to do with creating knowledge in our encounters with other places and other peoples. So I gave examples of different kinds contingency:

A Contingency of the **word**, or of the code, where investigation will reveal that even the most settled names for things were once a product of chance, or a mistake. To be aware of this is to give oneself the opportunity of coining new words for things as the occasion arises.

A contingency of the **event**, where in the case of a strange phenomenon like choreomania or a total eclipse of the sun, the conjunction of other contingent forces will conspire to make the event remain visible in the domain we call history where the ghosts of the past still speak to us.

A contingency of the **self**, where one's ongoing process of becoming takes on and incorporates contingent words and events and makes them part of a personal narrative.

A story which can be written, so there is also the contingency of **writing** which means developing a perception which persists like an after-image: Burroughs has his philosopher-pirate writing:

> Mission sat at the wooden table beside his phantom, his Ghost, contemplating the mystery if the stone structure. Who could have built it?
>
> *Who?*
>
> He poses the question in hieroglyphs…a feather…He chooses a quill pen. Water…the clear water under the pier. A book…an old illustrated book with gilt edges. *The Ghost Lemurs of Madagascar.* Feather…a gull diving for garbage…the wakes of many ships in many places (Burroughs 1995: 12).

ENDNOTES

1. Connolly (1993) tells us that Nietzsche enjoined us to overcome rage against contingency (to resist the temptation to convert some contingencies into essentialities). "What alone can our teaching be? That no one *gives* a human being his qualities: not God, not society, not his parents, not his ancestors, not he himself…He is not the result of a special design, a will, a purpose."

    In arguing for the acceptance of contingency Connolly notes that it is constitutive of subjectivity: "One happens to have these parents rather than those, to be gay rather than straight, to be persistent rather than flexible, to be sensitive rather than affable. These various contingencies enter into the self; they help to compose it. But they do not automatically fit together: they are likely to collide with each other in various ways…" (Connolly 1993: 162–3) and for its role in politics: "A democratic politics provides the best way to incorporate the experience of contingency into public life" (Connolly 1993: 159).

2. See Ian Hacking's interesting discussion about malostentions—radical mistranslations which become conventions—arising in cross-cultural contact, and specifically about the name of the lemurian *indri,* in "Was there ever a Radical Mistranslation?" (Hacking 2002).

3. The charm to avert the evil eye [notes the editor Frazer]; it is called in Italian "the fig", *la fica* or *mano fica*.
4. See my "Choreomanias: Movements Through our Body" *Performance Research* 8/4 (December 2003) 6–10.

REFERENCES

Andrianjafy, Dr (1902) *Le Ramanenjana à Madagascar, Choréomanie d'Origine Palustre* Montpellier: Editions du Nouveau Montpellier Médicale.

Astuti, Rita (1995) "'The Vezo are not a kind of people': Identity, Difference, and 'Ethnicity' among a fishing people of western Madagascar" *American Ethnologist* 22/3 (August) 464–821.

Bloch, Maurice (1971) *Placing the Dead: Tombs, Ancestral Villages and Kinship Organization in Madagascar* New York: Seminar Press.

Burroughs, William S. (1995) *Ghost of Chance* London: Serpent's Tail.

Chakrabarty, Dipesh (2000) *Provincializing Europe: Postcolonial Thought and Historical Difference* Princeton: Princeton University Press.

Connolly, William E. (1993) *Political Theory and Modernity* Ithaca: Cornell University Press.

Gezon, Lisa "Of Shrimps and Spirit Possession" *American Anthropologist* 101: 58.

Hacking, Ian (1998) *Mad Travelers; Reflections on the Reality of Transient Mental Illness* Charlottesville: University Press of Virginia.

-------------- (2002) "Was there ever a Radical Mistranslation?" in *Historical Ontology* Cambridge: Harvard University Press.

Hall, Richard (1998) *Empires of the Monsoon: A History of the Indian Ocean and its Invaders* London: HarperCollins.

Latour, Bruno (1997) "Stenger's Shibboleth" foreword to Isabelle Stengers *Power and Invention, Situating Science* Minneapolis: University of Minnesota Press.

Muecke, Stephen (2003) "Choreomanias: Movements Through our Body" in *Performance Research* 8/4 (December) 6–10.

Ovid (1967) *Fasti* (transl. James Frazer) London: Heinemann.

Peers, Simon (1997) "William Ellis: Photography in Madagascar, 1853–65" *History of Photography* 21/1 (Spring) 23–31.

Rorty, Richard (1989) *Contingency, Irony and Solidarity* New York: Cambridge University Press.

Smith, Nicholas H. (1997) *Strong Hermeneutics: Contingency and Moral Identity* London: Routledge.

Stengers, Isabelle (2002) "A "Cosmo-Politics"—Risk, Hope, Change" in *Hope: New philosophies for Change* Mary Zournazi (ed.) Sydney: Pluto Press.

Stewart, Kathleen (1996) *A Space on the Side of the Road* Princeton: Princeton University Press.

Unaipon, David (2001) *Legendary Tales of the Australian Aborigines* Stephen Muecke and Adam Shoemaker (eds.) Melbourne: Melbourne University Press.

part **THREE**

[NEW TECHNOLOGIES AND THEIR DOUBLES]

[ANDREW MURPHIE]

"Brain-magic": Figures of the brain, technology and magic

In the early days of cyberculture, virtual reality technologies seemed to promise an engagement with the non-sensory, a trip inside—or at least via—a brain separated from the problems of the body. The brain was fantasised as a magical place through which we could tour the entire world—a world that was entirely under our control. These fantasies of enhanced mind have a prehistory that is well recounted in the work of Darren Tofts and Erik Davis.

To take just one instance, which both Tofts and Davis relate, Renaissance intellectual and neo-Platonist Giordano Bruno believed in a mechanistic cosmos in which "the astral forces that govern the outer world also operate within, and can be reproduced there to operate "a magico-mechanical memory" (Davis 1998: 202). Here there is a correspondence between the mechanical and the magical, or between "symbolic logic" and "the divine attributes of God", between the complex movement of the cosmos and the complex movements of memory in the mind. Yet even as these correspondences give powers over the cosmos to the mind (and by extension in contemporary terms to the brain) they undermine the divisions between mind and cosmos, mechanics and magic. In Frances Yates' well-known *Art of Memory* the correspondences of this magico-mechanical complexity find their contemporary setting in the development of computers. For example, Werner Künzel's computer language COBOL came out of the symbolic logic borrowed by Bruno from "the thirteenth century Catalan mystic Raymond Lull".

The development of these correspondences attempts to spatialise or technicise the mind, turning spirit into mathematics and symbolic logic. In doing so, such developments try to avoid the virtual complexity of networks—which tends to slip through such logics. The brain becomes the house of rational, symbolic calculation—and thus the house of a thought which can (eventually) be controlled down to the smallest detail. Yet at same time, the attempt itself demonstrates that cognition remains central to what has famously been termed "Techgnosis" by Erik Davis.[1] This is a "term that recognizes the intersection of cyberculture and ancient, occult magic—'the expansion of consciousness by whatever means necessary'" (Tofts 1997: 81).[2] This expansion is not just a matter of symbolic processing. It is also a matter of networking.

In cyberculture the network complicates the idea of the expansion of consciousness (and processing of all kinds) with the complexity of the interactive world, with "imploded singularities unified around the concepts of immersion, interactivity and navigation" (Tofts: 74). It is in the network—not symbolic processing—that we perhaps see the technical mirroring of the complexity of the world most profoundly. When considered as a (perhaps *the*) crucial component of cognition, the network challenges "anti-magical" views of cognition and its place in the world. This challenge arises from the echoing of magic (as work upon the incorporeal, the distributed and unlocatable, the complex) within the technologies that assist (or create new forms of) cognition. I begin to approach this challenge by reviewing some recent reconsiderations of figures of the brain, technology and society in the light of networked complexity.

The cultural theory of the brain which emerges will also require us to rethink the relations between a techno-materialist culture and magic, ritual and force. In the process, we might be able to accept the role that metaphysical thought plays in culture (no longer relying on a metaphysics of the brain hidden within a materialist denial of metaphysics).

MAGIC AS FORCE/TECHNE

Let us define magic for these purposes. Generally, magic will be considered as the force of transformation, as active participation in the unknown (if not immaterial), or as the intensity of social actions that mediate different aspects of the material world (in particular the known and the unknown, that which we can control and that in which we have no choice but participation, despite our lack of control). To put this a little differently, I am positing a theory of magic as based on the presupposition that we have to work—everyday—with the unknown, at the junction of the known and the unknown, or simply with perceptions about which are constantly unsure. In short, magic not only accepts the unknown, it celebrates play or work with it. On the other hand, most forms of what I call here non-magical practice only predicate action upon what is capable of being predicted or known. Because magic allows more scope for action, it is more able to marshal forces in the world.

At this point, I also suggest that in some sense at least the common connection of technology to magic in general culture is a valid one. Magic has always been about power—over life and death and illness, over transformation, over appearance and disappearance. This is what technology is increasingly about as well. Although the discourses surrounding technology and the sciences are often derogatory when it comes to magic and metaphysics, we should be suspicious about this when they seek to take over exactly the powers previously ascribed to these areas. In short, all these terms need to be conceived in terms of the forces they marshal rather than their enduring claims on absolute truth.

Before considering magic as the marshalling of forces, we perhaps need to reconsider the dangers of reification in this area and indeed the notion of reification itself (in all its ambiguity). To do this, we turn to Michael Taussig's exploration of this question in his book *The Nervous System*.

SPIRITUAL FLIP-FLOPPING

In *The Nervous System*, Taussig first argues for the importance of not reifying social and other forces, not turning forces and relations into "natural things" (83). In relation to the signs and symptoms of disease, for example, to deny "the human relations embodied in symptoms, signs and therapy" is to "not only...mystify them" but to "reproduce a political ideology in the guise of a science of 'apparently' 'real things'—biological and physical thinghood" (1992: 84). For Taussig:

> ...medical practice inevitably produces grotesque mystifications in which we all
> flounder, grasping ever more pitifully for security in a man-made world which we

see not as social, not as human, not as historical, but as a world of a priori objects beholden only to their own forces and laws…(89).

Uncertainty and relationality, the very motive forces of cognition, are excluded in favour of mystification and absolute knowledge on one side (for the medical system), psychological pain on the other (for the patient). And as Taussig points out, what is precisely lost in such a medical approach is *the singular cognitive process of the patient*—which here includes the patient's own memory, experiences and the sensual/affective relation to others and the environment. It could be suggested that what is often lost in the hospital is just as often the focus of magic and ritual.

Yet this is to assume that the hospital is able to keep ritual at bay. This is far from the case. In the hospital we have what is best described not as science versus ritual but as an ultimately undecidable contest of ritual activities designed to bring different events/contexts into being (the rituals of medical science, the rituals of the singular cognitive process of the patient). This is not necessarily to valorise magic and ritual in themselves as politically radical. They can just as easily reinforce given social processes. Taussig notes Lévi-Strauss' idea that "the rites of healing readapt society to predefined problems through the medium of the patient; this process rejuvenates and even elaborates the society's essential axioms" (1992: 109). That these essential axioms now include the cognitive has been noted by many from Lyotard to Taussig himself.

Such contests are not restricted to the application of "hard" sciences. They are also common in social sciences. Taussig acknowledges all this and moves on to the more "provocative" political conception of a "flip-flop from spirit to thing and back again" (1992: 5). This, for Taussig, is "where the action" is, where what he calls a general social Nervous System "was put into gear, was in between, zig-zagging back and forth in the death-space where phantom and object stare each other down". In terms of the discussion about the brain as thing or species of event, we are talking about a theory that mobilises both conceptions. It also addresses mobility on its own terms.

It seems to me that without this mobility between dynamic networks and the singularities that form within them, we will never get close to a cultural theory of the brain. There is a sense in which the brain is both the most contingent of processes and also the "thing of things"—and the "thing" which justifies other forms of haziness such as the perception of the Other outside my head (the body, the other person) as a lesser "thing". The brain can be seen then as the literal embodiment of the entire problematic of reification—that is, in Taussig's sense of turning forces and relations into "natural things"—one that the magic of medical science is charged with solving for us. And it is properly a matter of medical *magic* rather than medical *knowledge* since, it could be suggested that magic—as the practical question of how to both acknowledge and deal with forces beyond our ken—is, rather more than knowledge itself, the theme of our age (thus once again suggesting the passing of the "information age").

Yet the reality of magic as social, forceful and thematic transformation is a *hidden* theme. It is hidden first because it is not really a theme—but rather the player and reorganiser of themes. Even considered as material process, magic is esoteric. It is hidden, more mundanely, because in its contemporary material ritual practice (including that of sciences, social sciences, the state and capital) magic is necessarily more effective if cloaked behind fixed "realities" and given knowledges—or at the least given methodologies. Even structuralist, deconstructive and critical *methods* have something to answer to here. Stephen Muecke insists, for example, on the falseness of academic "negative theology" (1999: 9) (which we could find in some structuralist or deconstructive methods) and "intellectual detachment" (which we could find in critical methods). For Muecke, "no hermeneutic tradition" or "ritual…has to be taken on its own terms" (1999: 10). Rather, such tradition is inter alia—"it means what it does via immediate relations between objects, things, feelings, words, music". How these relations play out is—in large measure—a matter of the cultural performance of magic.

Muecke writes that the cultural performance of magic occurs "in ways structuralists could never have imagined" (1999: 2). For Muecke, this magic is as much a question of "inflected" forces as codes, as much about the forcing of codes as the coding of forces. Moreover it is not so much a matter of "*formal*, textual, transformations beloved of structuralists". Rather, it is about "cultures *which work to enhance life forms*" (my emphasis). In short, magic is about life—the multiply-forced and coded life that exceeds the text. Magic is the amalgamation of forces and signs so that forceful acts of transformation and organisation take place. The sign's function within this is not the communication of meaning but its work with other forces. This means that "the presence of the sign is not an identity but an envelopment of difference, of a multiplicity of actions, materials and levels" (Massumi in Muecke 1999: 8–9).

As all culture continues to work "to enhance life forms" (for better or worse), we should not be surprised to find, with Muecke, that "'primitive' magico-religious forces are at the heart of nation-forming ceremonies [and elsewhere] in contemporary state society". For example, Muecke writes that the "state, relentlessly secular in its definition, remains none the less the highest form of the sacred-in-death" (1999: 3). This not only throws light on the ambiguous cultural status of the figure of the brain, but also on the brain in relation to the state—the brain as relentlessly secular, materialist and also the very metaphysical home of the forces by which state materialism is convened (not to mention the brain as definer of life and death). Muecke, writing about both Indigenous Australian and European state rituals, suggests that some of the "crucial relationships" here are between "body and country" (1999: 5).

I follow Muecke to José Gil's ideas of magic and force, as the arguments of both are crucial to an understanding of "brain-magic".

"BRAIN-MAGIC"

For both Muecke and Gil magic is real—materially real—insofar as it is a reorganisation of forces and energies (and therefore of bodies and signs). It is the focus in Gil's work on "practical effects", the "forces" that symbols can "draw on or shore up", and the "mechanisms…likely to trigger certain effects" that attracts Muecke to Gil. Gil writes that "it is not a question of studying forces (magical, religious, prestigious or whatever) according to their representational contexts, but to grasp them in the way they function in their own right" (in Muecke 1999: 11) within, and in the creation of, what we might call "force- fields".

The first of these magical force-fields is the body. Gil, using the term "exfoliation", describes the way in which "the body opens into the spaces it can occupy or articulate with" (in Muecke 1999: 13). Through exfoliations the body is "diversified" as a "volume in a perpetual state of disintegration and reconstitution". Although it is of prime importance to our relations within the world, especially to the relations of forces that make up cognition, exfoliation "only really makes itself visible in pathological or magical experiences" (Gil in Muecke 1999: 13). (Once again, the "hidden theme is the hidden theme".) Magic—whether used by the state or to escape the state/a state—is important here because its ritual is concerned with bringing about the extraordinary as a kind of socially organised pathological experience.

The extraordinary allows a more dramatic reorganization of relations within space than that presented by everyday life. In the ritual of the extraordinary, exfoliations occur that allow the body to create relations with forces in space in a transformational manner. And it is important to note that we are not dealing here with a "unitary body driven by a total self-image or central motor" (1999: 14). These ritual exfoliations, even if they present themselves as unitary, are never quite so. Neither are they ever innocent. Rather they could be seen as a basis for the subsequent ethical evaluation of transformations in culture and the possibilities of experience.

I would suggest that Gil's account of magic and ritual, signs and forces, allows us to consider the brain as one figure—perhaps *the very* figure—of magical transformations of forces in the West. In conservative transformations, the figure of the brain gives a sacred (hidden and metaphysical) underpinning to the ongoing reformation of states (from the deployment of pharmaceuticals in the cognitive setting to ritual performativity in the workplace). This perhaps explains both the proliferation of *disciplines* surrounding the brain and the political need to work between these disciplines, to break them down.

In the contemporary world these disciplines are indeed beginning to talk to each other. When they talk about the emergence of cognition "via immediate relations between objects, things, feelings, words, music" (Muecke 1999: 10), they might begin to ask questions about the "brain-magic" towards which we have been heading throughout this essay. We could ask them, not about work on improving symbolic processing, but about the "envelopment of difference, of a multiplicity of actions, materials and levels". (These questions are of course increasingly central to research into cognition within connectionist and dynamicist approaches). We could ask about the various topologies of the brain—whether true, false or half-true—and their organisation and transformation of forces in the world. We could see the brain *as body*, and thus as exfoliating (including, as with the body, a series of diverse "infoliations"). The exfoliation of the brain in space would occur both with regard to the rest of the body (and nervous system) and with regard to the world, as best summed up perhaps in dynamicist theories of cognition such as Andy Clark's notion of "extended mind". Such questions would suggest once again that, as with the body in the midst of forces, we are not dealing with the brain as either central motor or as a unity. Indeed, as Taussig writes about the nervous system as a whole:

> Even while it inspires confidence in the physical centerfold of our worldly existence—at least that such a centerfold truly exists—and as such bespeaks control, hierarchy, and intelligence—it is also (and this is the damnedest thing) somewhat unsettling to be centered on something so fragile, so determinedly other, so nervous (1992: 2).

"EXTENDED MIND"

Some contemporary conceptions of thought as more and more dispersed, make this an increasingly fragile "centerfold". Andy Clark, in his convincing argument for the significance of embodiment to the brain, and for what he describes as "extended mind", compares the movement of thought through the world to that of tuna through the water. These fish appear, like many others, to be able to generate more speed and power than their muscular structure should allow, primarily because they ride the eddies and vortices they create with their own tails. As Clark writes:

> Ships and submarines reap no such benefits: they treat the aquatic environment as an obstacle to be negotiated...[while] tuna...profit profoundly from local environmental structure...This simple observation has, as we have seen, some far-reaching consequences...gone is the neat boundary between the thinker (the bodiless engine) and the thinker's world...it may for some purposes be wise to consider the intelligent system as a spatio-temporally extended process not limited by the tenuous envelope of skin and skull. Less dramatically, the traditional divisions among perception, cognition, and action look increasingly unhelpful (1997: 219–21).

Extended mind, extended and distributed brains, thought as distribution—and none of these given once and for all. They are all in an active co-emergent series of what Varela et al call "structural couplings", but which are obviously not limited to "couplings".

One consequence of this interactive activity is that many authors point to the importance of "technique" within judgement and thought (Connolly 2002; Clark

1997). Technique involves the manner in which we bring together distributions and structural couplings, transform them, pull them apart, work through them or, better, learn how to participate in them. Marcel Mauss (1992) noted that anything to do with the use of our bodies involves technique, as long as this technique was effective and traditional (it could be passed on), with the result that "we are everywhere faced with physio-psycho-sociological assemblages of series of actions" (Mauss 1992: 473). Thought is of course immersed in these assemblages, with the brain as part of the technical body, and the local world as extension of both. Indeed, the question of technique within thought now arises with some urgency. In general, it is not just that thought is revealing itself more and more to be based upon a series of techniques. It is also that it needs more of these techniques, and invents more and more of them to negotiate fragile assemblages of cognition, thought and worlds, conscious and unconscious. Perhaps another return to something like Freud's unconscious might now be occurring via Clark's tuna and their vortices (although this is not an unconscious created by repression as conceived by Freud). It appears that the gap (or threshold) between conscious and unconscious, or between consciousness and world, is again turning out to be the most profitable ground for thinking about thought, even for directing thought via technologies and techniques. Connolly (1999), for example, takes a wonderful turn on Kant, based on the half-second delay between sense and consciousness. He wonders whether this implies a re-situation of Kant's transcendental "supersensible domain…in the corporealization of culture and the culturization of corporeality". He proposes an "immanent naturalism…in which the transcendental is translated into an immanent field that mixes nature and culture".

Again there appears to be an immanent virtuality to this that is necessary to even begin to understand the basics of thought. For Connolly, this is because thinking is "irreducible to any of the ingredients that enable it, but is also affected profoundly by the infrasensible media of its occurrence". In other words, thought is embodied interaction and when interaction is primary, we are in the realm of the virtual. Connolly's project in his recent *Neuropolitics* is precisely to point to techniques that can both acknowledge and work with this fullness and excess. As shall be seen, it is here that I would locate the reality—and technics—of magic and ritual. We may need magic and ritual more than we acknowledge within the cultures of cognition.

BRAIN ETHICS

Let us, then, tentatively extend Gil's ideas about the body to the brain—the brain conceived as "interval" (Bergson) or "screen" (Deleuze). If the brain is a host of magic in any real sense, it is precisely such a host by virtue of becoming a component of the body's exchange of codes and energies without itself signifying. To put this in other terms, the brain perhaps signifies so much within culture precisely because it does not work through signification itself. It works not as a storer and regulator or recaller (in memory) of "held" forces, or as a calculator of systems, but as empty active exchange of energies. As Gil writes of the body, the brain "does not speak, it makes speech...it provides language with a virtual and silent 'grammar'" (1998: 111).

This very real "brain-magic" gives itself to what we might call an "ethics" of the brain. It allows for new powers related to the brain. It also explains current powers of brain-magic that arise when understandings of the brain are turned into disciplining sciences that in turn facilitate the sciences of attention, distraction and cognitive ergonomics. There is much as stake here. For a start, the more that is claimed for the central powers of the brain, or for what might be "held" within the brain, the more the brain seems subjected to pre-arranged powers (as opposed to *participation* within them). In this regard, Gil writes of a body not allowed any play in the release of coded energies. This is a body subjected to "state magic"—to a space of pre-coded energies and spaces or topologies.

The already organised space, the discourses, the ritual sequences submit the body to a spatial discipline. It does not translate codes, codes are translated in it; it does not exfoliate, exfoliations are already given in space. Ritual action, on this level, consists in forcing the body to go from one space to another, to follow a translation already realised in myth and space (Gil 1998: 119).

Such a reactionary deployment of ritual is, perhaps sometimes despite itself, still creative (and unpredictable). Indeed, even in this context, we can understand the function of magic as the translation of forces and energies *in creating the very notion of magic, or the sacred itself*. For Gil, this involves the production of a "'sacred' or 'magic' unconscious" (1998: 142). Within "state magic" this is an unconscious in which the hidden theme is created *as* the hidden theme—in which operational magic and the sacred are not only repressed by discipline, but *produced as the repressed of the state* at the same time. In the regime of the disciplinary, one must not only mind one's thoughts, but the very conception and topologies "wherein" they are supposed to take place. Which is to say that one must (impossibly) reject any singular access to the virtual and the affective precisely as discipline requires. In state magic, at best, "the affective energy, for which the myth is that ultimate metaphor, must be reinvested on the body of the adept" (Gil 1998: 143) in the appropriate manner. As Gil points out, the prime aim of "symbolic efficiency" is the effective "remote control" of affect and the forces involved.

This is an argument for the crucial work of affect (here considered as the movement of forces and so on, their impact upon each other in time and space), even within the realm of the brain. Such ideas can also be found in the more dynamicist side of the cognitive sciences and philosophies. Yet Gil's understanding of magic has a great deal to contribute as well not only to the cognitive sciences but to any disciplines which assume cognitive processes as foundational to disciplinarity. An example is media studies. Here media is conceived as containing some kind of science of communications (and not only within media departments in universities but within the whole world of media, which is awash with assumptions about cognition, message, producers and receivers, symbolic processing and affect…). As Gil writes, all such "sciences" are always already caught up in the tensions involved:

> …the first pathways of science are traced in a permanent tension and amalgam between old resurgences of magical and religious thought and the logic belonging to the new requirements of rationality and experimentation (1998: 149).

Where does this tension leave us? Perhaps arguing for a mix of post-connectionist, dynamicist understandings of the brain and an anthropology or ethology of the techno-social and other ecologies within which the brain works. These are ecologies in which affect and the heterogeneity of the world (or worlds) bring us back to the importance of magic as work with forces.

To put this differently, we are not our brains, as we often assume or are told. We are rather the rhythms of the world that our brains only partially modulate with their own rhythms. This makes the brain a somewhat delicate assemblage, one we can admire the more for its incompleteness and for its magic. Moreover, this suggests that there should be a delicacy of ethics surrounding the brain, something that might encourage us to think carefully about the fragility and uniqueness of our temporal and ethical sites of immersion in the world. These positions again call for magic if by this we mean the work of transformation and connection.

In briefly reconsidering the ethics of networks surrounding cognition, we have reconsidered the ethics of the brain and the transformational and connective assemblages in which we find it. Networks are an attack on the very conceptual foundation of older media, and with these, older forms of social organisation based upon representation. They also call out for a subtler, pragmatic and more ethical approach to a world that comes before, will come after, often perhaps despite representations and stable frameworks. Virilio has referred to this complex as the virtuality of images (1994: 59), and often laments the emergence of this virtual (and technical) complexity in the place of the images we once knew. His criticisms should be taken seriously, yet not all is lost. It could be argued that the virtual complexity of the network—especially that within and around the brain—takes the senses more seriously than ever before. We need to do more than just posit their loss in the face of the worlds that new technologies make us begin to think.

ENDNOTES

1. Of course, this is the primary text about the indissoluble binding between magic and technology although other important texts are David Noble's *The Religion of Technology* and Jeffrey Sconce's *Haunted Media*. All of these texts attest to the fact, that far from being the enemy of science, magic is the mediator between science and its other (whatever that other may be). And far from being the enemy of the spirit, technology is the mediator between spirit and its other (which in this case often seems to be science). From a cultural analysis point of view, all these books attest to something like Bruno Latour's actor-network theory.
2. And here also, lest we forget, the task in understanding the brain, cognition, and its relation to the world, and to technology, is not just "discovering it" through brain scans and so on. We are always constructing 'the brain' as we go, and this is a process that shall perhaps remain unfinished, as new archive after new archive emerges from the virtuality of the archival itself (thus the opposition between those such as Daniel Dennett (1991) who claim to "explain" consciousness minus a few details, and those such as neuroscientist Susan Greenfield, who though they enjoy the chase, often profess a doubt as to whether we will ever "understand" our processes of understanding).

REFERENCES

Clark, Andy (1997) *Being There: Putting Brain, Body and World Together Again* Cambridge, Mass.: MIT Press.

Connolly, William E. (1999) "Brain Waves, transcendental fields and techniques of thought" *Radical Philosophy* 94 (March/April) accessed on-line at http://www.radicalphilosophy.com/default.asp?channel_id=2188&editorial_id=10185 on 10 May 2003.

-------------- (2002) *Neuropolitics: Thinking, Culture, Speed* Minneapolis: University of Minnesota Press.

Davis, Erik (1998) *Techgnosis* New York: Three Rivers Press.

Dennett, Daniel (1991) *Consciousness Explained* Boston: Little, Brown and Company.

Gil, José (1998) *Metamorphoses of the Body* Minneapolis: University of Minnesota Press.

Greenfield, Susan (2000) *Brain Story* London: BBC.

Latour, Bruno (1999) "On Recalling ANT" in Law and Hassard (eds) *Actor Network Theory and After* London: Blackwell.

Lyotard, Jean-François (1984) *The Postmodern Condition: A Report on Knowledge* Minneapolis: University of Minnesota.

Mauss, Marcel (1992) "Techniques of the Body" in Jonathan Crary and Sanford Kwinter (eds.) *Incorporations* New York: Zone 455–77.

Muecke, Stephen (1999) "Travelling the Subterranean River of Blood: Philosophy and Magic in Cultural Studies" in *Cultural Studies* 13/1:1–17.

Noble, David (1997) *The Religion of Technology: the divinity of man and the spirit of invention* New York: A. A. Knopf.

Sconce, Jeffrey (2000) *Haunted Media: Electronic Present from Telegraphy to Television* Durham: Duke University Press.

Taussig, Michael (1992) *The Nervous System* New York: Routledge.

Tofts, Darren and McKeich, Murray (1997) *Memory Trade* Sydney: Interface.

Varela, Franciso J., Evan Thompson, & Eleanor Rosch (1991) *The Embodied Mind: Cognitive Science and Human Experience* Cambridge, Mass.: MIT Press.

Virilio, Paul (1994) *The Vision Machine* Bloomington: Indiana University Press.

Whitehead, Alfred North (1978) *Process and Reality: An Essay in Cosmology* New York: Free Press.

Yates, Frances (1966) *The Art of Memory* London: Routledge and Kegan Paul.

[CHRIS CHESHER]

The muse and the electronic invocator

To search for relationships between magic, religion, science and technology is already to have begun too late. There is no reason to accept these as absolute and exclusive categories. This chapter shows that cultural practices need not respect boundaries between systems of ideas, media epochs, discourses or institutions, but recur across a wide range of contexts. Although the overarching systems of belief called magic, religion and science/technology are usually seen as incompatible, and often in direct conflict, they share similar aspirations, methods and histories. Each aspires to extend perception and action beyond ordinary limits, and all share similar habits and procedures of ritual and power.

A good example of a practice that persists across magical, religious, artistic, legal and scientific contexts is the invocation—a call to a power outside ordinary fields of perception and action for immediate assistance, guidance or support. This cultural form has a long and continuous history through changes in media, technology, institutions and discourses. Magicians invoke spirits; priests invoke the Name of Christ; artists invoke a Muse; lawyers invoke precedents; scientists invoke previous experiments, and natural laws; a computer program invokes a subroutine. The invocatory act is neither uniform nor universal, but it is a distinctive mode of ritualistic performance that returns in diverse cultural practices with remarkable regularity.

The persistence of invocation is particularly significant in contemporary life because it is embodied in the design of computers. Programmers have long used "invocation" as technical term, but the deeper appropriateness of this metaphor is not usually recognised. Rather than computation, it is invocation which makes computers distinctive: they call things up. This chapter traces the translations of the invocatory act through its manifestations in oral, literate, technical and modernist forms, through to its emergence in the now ubiquitous electronic invocator.

The invocatory act is in part characterised by distinctive structural features: a point of crisis; an existing relationship with something outside ordinary life; a voice that makes a call; a medium which carries that call through space; and codes or protocols which must be followed (avocations).[1] However, these features are not sufficient to secure the efficacy of the invocatory event itself. An invocation is always singular, and contingent on the conditions in which it is performed. An invocatory act is an intervention through speech (or an equivalent) that aspires to structure the unfolding of an event in the future. In Heidegger's terms, invocation relates to processes of *poieisis*, or revealing. Better, invocations speak to future becomings. By speaking certain names and formulae, following ritualised procedures, a speaker (or equivalent) acts to reveal a state of affairs that responds to the current crisis or desire. This process calls into the virtual to summon processes of actualisation.

Invocation is one of the key mechanisms by which social power is established and sustained. Invocations mediate relationships between speakers and collectivities, and operate at the interface between present and future. They put an event at a particular time and place into a relationship with enduring, intersubjective cultural abstractions and social institutions. The invocatory act calls to an authority outside ordinary life: an authority accessible only by virtue of the speaker's own special position. A king invokes divine authority by claiming special relationship with the gods; a judge invokes the authority of the court and the law; users invoke databases to read or change information, but only within the limits of their access privileges. An individual's invocation seeks immediate authority, guidance or inspiration but, in doing so, also affirms existing power relations.

But not all invocations are the same. Magicians call on natural occult forces for immediate action. Priests and believers in religions, on the other hand, tend to build faithful relationships with deities (and with religious institutions) through prayer. Fortunate events may be deemed to have been miraculous. When a prayer is answered, this is quite unlike the almost mechanical efficacy expected of magical spells (Thomas 1971). Invocations can be desperate and futile cries from powerless voices, or irresistible commands of power. They can be closed or open: overdetermined commands, or open-ended appeals. They can be spoken, written or computerised. They can involve all manner of combinations of humans and non-humans. Some invocations are generally respected, while others are disdained. Some invocations are explicit, but the most powerful are implicit: being executed and immediately erased. Invocation is often but not always gendered: the invoking voice is by default masculine, and the invoked entity is typically feminine such as the Muses or the blindfolded figure of justice.

Showing that there are cultural continuities such as the invocation undermines the myth that there are complete revolutions in human affairs. If it were true that the present is unprecedented, modernity could be absolved from being judged alongside all the mystifications of the past. However, the persistence of invocation shows that power and language continue to have metaphysical dimensions.

INVOCATION IN GREEK MYTH

Greek mythology established the Western archetype for the invocation. Invocation is the main way mortals communicate with a pantheon of gods that reign over the ancient magical imaginary. The distinctive elements of the cultural practice of invocation are already apparent— a crisis, a relationship to an outside Other, a voice, a medium, and a protocol. When the mythical hero in Homer's *Odyssey* loses control over his own destiny in a crisis (in this case a shipwreck), he speaks aloud to invoke a convenient god for immediate assistance:

> Hear me, lord, whoever you are, I've come to you, the answer to all my prayers— rescue me from the sea, the Sea-lord's curse!…Pity me lord, your suppliant cries for help (Fagles 1996: 166 [*Odyssey*, Book 5, 490–7]).

Odysseus' invocation is based on an ongoing, but unequal, relationship with the gods who, unlike the drowning man, are unbound by space or time. They can move anywhere, and see anything. They can perform transformations, create illusions, hide and reveal things, and make prophecies. By contrast, mortals are subject to gods' unpredictable vindictiveness and generosity. Poseidon caused the storm, but the goddess Pallas Athena rescues him from it. Within this mythical world, invocations assert a brutal but regulated divine order. In making invocations, mortals speak as suppliants. Their entreaties reverberate through a celestial hierarchy. They seek immediate power, but in following protocols and repeating the gods' names in reverential terms, re-affirm the hierarchy. In receiving good fortune, the hero also accepts his humble position in relation to the forces of destiny.

The oral culture of pre-Homeric Ancient Greece reverberates with invocations at many levels, not only within the fictions of the stories. The stories themselves are invoked. The storyteller does not claim authorship. Rather, he says that a goddess inspires him—the Muse literally gives him breath. The Muses are nine minor goddesses whose job is to protect the cultural heritage of Greece. They personify the traditions of the community within specially defined areas: history (Clio); astronomy and astrology (Urania); epic poetry (Calliope); dancing (Terpsicore); song, rhetoric and geometry (Polyhymnia); comedy (Thalia); tragedy (Melpomene); lyric poetry (Euterpe); and love poetry (Erato). Members of the community can identify with the vocations associated with each of the goddesses. An artist's individual authority is based on claiming a special position in relation to a particular Muse. The storytellers have both power and responsibility. Their places as speakers are conditional on telling the story well. The gendering of the relationship between storytellers and Muses is itself a significant part of this patriarchal heritage—a mythical regime where the feminine force is the creative wellspring, but also silent and out of reach.

The motivation to tell a story, and the choice of which story and how to tell it, is always a response to some form of *crisis*, whether it is something urgent (a natural disaster), or mundane (boredom). The storyteller answers calls for reassurance, entertainment, education or *amusement*. Unlike individual desires or wishes, which are silent and private, invocations are spoken, and public. They are not responses to individual needs, but affirmations of collective identity. Invocations express desires for knowing one's place in the world. Any answer will give invokers a sense that they are part of something larger than themselves.

Telling a story uses words to call into presence a world that is not ordinarily apparent. The storyteller's voice evokes a whole universe, built up with description and narration to bring to mind scenes, characters, events and moral forces. In oral storytelling, tales are not recounted verbatim, but genuinely invoked from memory and wit— they are never the same twice.

However, invocations are not conjured from nowhere, but follow strictly prescribed protocols. Each telling is built from common formulaic components— the oral cultural form of what I refer to as "avocations". Stories are built up from standard elements: a prescribed domain of names (gods, heroes or other archetypical characters), a set of operations that can be performed (escape, brave deeds, conquering towns with clever tricks, magical forces, navigating across the sea); and formulaic fragments of language—"brave Odysseus"; "generous King Ancinous"; "lovely Polycaste". These and other mnemonic tricks help storytellers maintain consistency and keep the flow in real time. Even the listeners often already know the stories. When invoking a Muse the storyteller is actually invoking his memory, his training, and the collective memory of the entire community (Ong 1982).

The power of stories in oral cultures does not stop with mythical allegories, but extends into other forms of knowledge and practice. The tradition of alchemy, practised in China, India and Egypt even before Ancient Greece, performs proto-scientific, or magical invocations that purport to give the speaker control over natural forces. Alchemists typically aim to create gold from other metals, or to produce elixirs of immortality. The practice of alchemy involves gathering special materials (metals, medicinal plants, bodily fluids), and subjecting them to careful procedures, as well as performing ritualistic meditation and incantations, often in specially charmed places. It is obscure which of the operations are effective manipulations of physical phenomena, and which are superstitious conventions. The invocations call both to spiritual forces, and to physical properties in the materials. Scientific analysis might some time show that some of these rituals produce actual chemical reactions, while others rely on the gullibility of participants, but that is no doubt true of many scientific practices as well. Final judgements about truth and illusion can remain indeterminate without compromising the efficacy of invocatory events, which operate through singular combinations of language, politics, technology and artifice.

INVOCATION IN WRITING

The cultural form of invocation does not disappear with the decline in prominence of oral storytellers and alchemists, but is translated to suit the available new media. In spite of transitions from "primary oral" culture to "chirographic" and "typographic" culture (Ong 1982), the practice of invocation continues. Its re-mediated forms are indeed different. With the aid of writing, it becomes possible to repeat invocations perfectly and indefinitely. They can be tested for logical and factual consistency and can be extended in space and time, independent of individual memory. In writing, spoken invocations become citations, by which authors refer to the text of other authors to give their own arguments more authority.

Written invocatory acts replace a speaker's voice with the imagined voice of an absent author. The unheard authorial voice is only ever incomplete evidence of a hidden truth in the original spoken voice. This invocation of presence in a text promotes a systematic privileging of the authority of the text, or logocentrism (Derrida 1976). Any actual reader's voice (aloud or silently read) is only a distant copy of the original statement from the author. The authority of any text emerges from this invocatory gap between the true voice of the author and the imperfect voice of the reader.

A spoken utterance is said once, and disappears then into the vagaries of memory. A written statement abstracts the invocation from a singular temporal context. The invocation becomes virtual in a different way because avocations become material. In oral cultures it is human memory, memory aids, learnt techniques and material artefacts that function as avocations. Writing physically inscribes the avocations, and detaches them from a human speaker. Written text can be read by anyone familiar with the conventions of the script. The avocation becomes invocable in multiple contexts to anyone with access to the text. With writing, invocatory protocols could become more entrenched and more intricately developed. Plato's *Republic*, for example, develops a totalising utopian system of named social roles, attributes and abstractions (Plato 1974). His writing provides a model followed in much future political thought. The mandatory invocation of Plato, whether embracing or rejecting him, provides some of the most influential vocational infrastructures for Western political theory.

Text itself increasingly becomes a primary source of authority. Where spoken invocations call on the authority of a charismatic individual, or a higher

Age of Myth and Legend (1993) by stressing that his work makes no claims to useful knowledge, but that "literature is one of the best allies of virtue and promoters of happiness" (1993: ix). Bulfinch's account of Greek and Roman myth provides a systematic cataloguing of ancient myths. He seems to be motivated by a commitment to the value of preserving this knowledge, but at the same time, distances the practices he reports on as definitely foreign and other. Classical knowledge is presented in diagrammatic form, with glossaries and genealogical tables, and observations about the logical inconsistencies implicit in the mythological ontology. These modes of invocation are appropriate to the domains of artistic expression (especially in poetry) and historical curiosity, but have no place in a demystified modern era.

Meanwhile, the emergence of the technological, discursive and political hybrids that Latour identifies include a proliferation of devices that perform invocations. Powerful invocatory devices appear first in industrial, and later domestic settings. They appear at an accelerating rate during the twentieth century. All manner of material and symbolic effects become available "at the push of a button".

INVOCATORY TECHNOLOGIES

Speech is conventionally deemed to belong to the category of culture, and therefore something other than technology. However, it is apparent that many devices perform physical or symbolic actions which are equivalent to speaking. For example, when a doorknocker or a doorbell is installed on the front door of someone's home, it stands at the ready to make a statement or a request when someone comes to visit. A knock is a demand, with both meaning and force—it operates both at the level of meaning, and action. The knocker functions as a substitute for speech, and as a pair to the gesture (of knocking). It channels visitors towards a particular course of action—knocking (rather than screaming out, banging on the door with a fist, or breaking in through the window). Therefore, it is not simply calling for the owner's attention, but also invokes higher order cultural values of hospitality. The knocker is an avocation for the subject position of the polite visitor, and encourages the visitor to follow the conventions of the visit. A knock on a door calls to the resident, first, but it also invokes the social conventions about how to receive guests.

Many technologies have an invocatory dimension—they provide a mechanism by which a user can attempt to call up a predictable effect, as if they were speaking to something outside the ordinary lifeworld. The most rudimentary of technologies—the lever—is a very abstract invocatory device. It functions paralinguistically, where phrasing is apparent in the physical arrangement of fulcrum, resistance arm and effort arm. The lever divides objects into a particular relation to each other (load and force) in the same way that language divides the world into objects and subjects. Technology and language are traditionally considered separately but this example shows that they are in many ways indistinguishable. The diagram of the lever is a phrase-form that can be invoked in many avocational assemblages. The lever form is invoked in the construction of doorknockers, seesaws, catapults and pistols.

Any actual invocatory device (as opposed to the virtual principle of the lever) always involves more than a material effect. When invocations are translated through technological systems, there tends to be coalescence between invocational voice, instrumental force and operational rituals. The invocatory device typically black-boxes the invocation within the device's structure, and the conventions about how it is used. For example, a pistol may be invoked to back up someone's position in a crisis. Its parts have an unambiguous phraseology: the trigger is hidden behind a guard, close to the index finger of the holder. The deadly force summoned by pulling the trigger makes little sense on its own (it provides force but little meaning). However, its action is always accompanied by a repertoire of invocable justifications. Concepts such as self-defence, lawful killing, or the state of war are just as significant to the successful operation of the weapon as the gunpowder. There are also rituals associated with the device: training in handling guns uses incantations to reinforce certain appropriate rituals—always hand over a gun unloaded; never point the gun at someone you do not ntend to shoot; check three times the gun is unloaded; then check again. These may not be magical superstitions, but they operate in the same way, associating the object with affects of fear and power, and inculcating the user with routines that become unconscious.

Many modern technologies have an invocatory dimension (but not *everything* is an invocation). The light switch calls on light; the radio dial calls up a radio station; dialling a telephone number invokes the network to connect them to someone elsewhere. The illumination, the radio program and the telephone conversation are invoked only once, and as long as they continue to be effective, they are not invoked again. The crisis that provoked the invocation (darkness;

silence; loneliness) may be partially resolved by the actions of the device itself. The invocation itself may be quite trivial. However, if it fails, that will tend to draw some attention (blackout; signal jamming; a telephone out of order). The availability of invocatory devices is often taken for granted, and there is some validity to Heidegger's warning that dependence on invocable standing reserves systematically hides other possibilities (Heidegger 1977). When invocatory devices are installed, a domesticated version of the crisis is built into the invocational infrastructures, so that a failure of the system produces a new order of crisis: a blacked out city has greater dangers than a city accustomed to the dark.

As well as devices that offer users the capacity to invoke, there are invocatory devices that also discharge invocations. Traffic lights effectively regulate traffic flows by implicitly invoking road rules. The protocols by which they operate are technical and political—electronic and legislative. They work partly on the basis that transgressions may be punished, but more through a respect for the authority of the inhuman signal. Red light means stop. This is not a meaning, but a command. Traffic lights were only one of many specialised invocatory devices which became familiar across the cultural landscapes of industrialised societies. However, these early simple invocatory switching technologies would be seen as only progenitors to the general purpose electronic invocator.

THE ELECTRONIC INVOCATOR

By the middle of the twentieth century processes of mass urbanisation, a proliferation of arbitrarily connected invocatory devices, and an expansion of colonial bureaucracies are tending towards a new order of crisis. The problems emerge from an explosion in the scale of human activities: massive populations; accelerating speeds of transportation and communication; and overextended chains of command. This crisis in control reaches its zenith in the Second World War when the principles and components for invocational media are first assembled.

In the midst of this crisis, a new mode of invocation appears in the form of a technological apparatus that internalises the invoking voice and puts invocations and avocations into the same automatic circuit. A surrogate-invoking subject (the central processing unit) is directly wired into a specialised collection of invocable domains (memory, logic and maths units and peripherals). It can remember, follow and copy complex instructions, and make decisions, performing work that was previously reserved for humans. The device is called the computer because it does the same work that computers had—human computers to whom scientists in large research facilities delegated their tediously labour-intensive calculations. In fact computer pioneer Von Neumann's design for the computer is itself an invocation of the human nervous system, as he outlines in the book, *The Computer and the Brain* (1958).

The term "computer" is problematic though since only a fraction of invocations performed are computations. Invocational protocols include far more than mathematics. For example, only a small subset of machine language commands for the x86 microprocessor in PCs are arithmetical. This processor recognises instructions to add (ADD); subtract (SUB); increment (INC) and compare (CMP)) values. However, the majority of invocations perform other functions: such as calling to memory (MOV; PUSH; POP), invoking input / output operations (OUT; IN), control transfer operations that make conditional jumps out of sequence (JZ; JGE; JB), and interrupts and exceptions. Many of these instructions bring peripheral devices into play—keyboards, controllers, screens and speakers—which are "dumb". But they are as important to the device as the CPU. Considering all these components and capacities, the term "computer" is inaccurate. These devices work by mediating a heterogeneous range of invocations, and should be called general-purpose electronic invocators.

Invocators generate and resolve millions of invocatory crises per second. They reduce the invocational voice to discrete voltage variations and magnetic pulses. They position images of the outside Other in invocable domains of memory, storage and networks. They strictly follow programs, standards and protocols. The result is a highly abstract medium that offers to call up almost anything. Invocators can be programmed and reprogrammed indefinitely to invoke new things. The results of one invocation can be incorporated into further invocations, in series of iterations. They work by repeating thousands or millions of simple programmed steps that invoke instructions from memory, interpret them, perform the operation, and move to the next instruction. This repetitious operation invisibly recalls not only Ford's industrial production line but also the ritualistic procedures of magical incantations.

Invocators are dramatically more versatile than any invocatory device. In some ways they have some of the same elements as a lever. Levers have special points of input and output around a fulcrum. The invocational assemblage also has specialised input devices (card readers; keyboards; mice) that compose and initiate invocations, and output devices (printers; screens) that express the manifestations of invocational processes. The CPU is the fulcrum for invocations, a set of circuits through which all signals are switched. However, its invocations are sequences of electrical differences, and not physical forces. Unlike the lever, where the action at the output is always simultaneous and proportional to the input (which is why it is called an analogue technology), there is always a temporal and physical interval between input and output. The signals from inputs are stored in memory devices and remain invocable as data and programmed avocations. Within the invocational interval, avocations invoke very complex and intricate digitally encoded multilayered environments: operating systems, special-purpose applications, and virtual spaces.

Invocators support intricately defined positions for users and extended avocational vocabularies. With a lever, the justifications for use, and the training in the techniques for using it are largely extrinsic to the device. By contrast, invocational systems can partially incorporate access restrictions and help systems alongside the material actions which they control. Password and security mechanisms determine who can pull the virtual lever, and in which directions. Of course the input "lever" does not have to take any particular physical form— any device that can measure an environmental difference can function as an input device to invoke events. An image on a video camera can invoke a facial recognition process; a keyboard provides an arrangement of keys that invite users to invoke things by typing. Many of the processes that users invoke are internal to the system —file management; help files; and simulations. Many of these processes are partly contractual—"Click OK to continue"; "Are you sure you want to do this?" The output mechanisms can be multiplied indefinitely: files can be printed out many times or can be distributed through networks.

Invocational media perform citations far more quickly, consistently and automatically than scriptural technologies. Networked invocators provide users with authorisation, guidance or special powers. They orchestrate complex arrangements of electronic voices to accelerate existing relationships with outside entities. Sometimes they invoke a Big Other (the state; the corporation). Large systems established by governments and financial institutions invoke internally consistent and interconnected invocable domains that give tangible presence to the kinds of economic and legal abstractions that McKenzie Wark refers to as "Third Nature" (Wark 1994). Third nature is the invoked overlay that operates on top of natural and social environments. Even if it is entirely conceptual, it often works with considerable force, dictating who should be incarcerated, fed or employed. Large-scale invocational systems mediate records about insurance, police matters, taxation, payroll, and so on. When individual subjects have dealings with institutions that manage these records, operators of invocators make queries to the system to establish immediately the status of this person. Equipped with networked databases, invocators call up, quantify and visualise abstractions such as credit, debt, risk, entitlements, criminal record, citizenship, and the like. Like the Greek pantheon, these invoked entities are not restricted by space. The invocator gives these abstractions an intersubjective existence independent of any human agency.

However, not all invocations call on a Big Other. Many invocations are effectively intransitive—they do not invoke anything specific. They might call up a playful game world or an artistic experiment. For users, there is often little apparent difference between playing a game and doing something "for real". The head-up display used in weapons systems resembles many computer games. Database technology handles billions of dollars or the police records of states but also mediate discussion forums or manage trivia. The power of invocation is not intrinsic to the technology itself. There is always something invoked that resides outside the materiality of the system, and must be collectively recited and believed. Fictions such as the "law"; the "economy"; and the "nation" are brought into existence only by being constantly invoked. However, the event of invocation itself tends to be fetishised, particularly when it is calling nothing! There is little mystery in a database of widgets unless the existence of widgets is somehow in doubt.

The general association of invocators with mystical power has generated a hyperbolic sense of the invocational uncanny. The powers of these devices often seem unsettling and supernatural. The interminable artificial intelligence (AI) debate rehearses the problem of looking for true Intelligence. The AI and robot enthusiast Hans Moravec dreams of invoking immortality by uploading his mind into invocable domains (1988) or invoking a superior robot species that will replace mere mortals (1998). The aura around AI research, which Dreyfus (1966) derides as alchemy, is generated by its aspiration to invoke something intangible. Advocates of virtual reality summon a similar power from their promise of a future system that will be indistinguishable from "real" reality. The rise of technopaganism in the 1990s (Rushkoff 1994), and its intermixing of retro-counterculture with technology to produce cyberdelia (Dery 1996) are refrains of far older cultural patterns. But these hybrids of high tech and magic are nothing exceptional. As Erik Davis (1998) shows, there has been a constant interplay between magic and information technology, from ancient alchemy to the extropian movement.

By tracing the persistence of the practice of invocation across all the supposedly untranslatable domains of magic, religion, politics and technology, I hope to stress that there are continuities through change, and that historical and discursive boundaries between modes of knowledge are quite artificial. Invocation begins by defining a crisis. It secures a relationship to something outside ordinary life. It articulates a call with a voice through some medium. It follows and thereby defines or reinforces a set of protocols. In itself, invocation carries no inherent moral value. It is a powerful practice that comes into play in inciting cultural cohesion and change. Every invocation, whether mediated by the human voice, written text, or the electronic invocator, calls on a metaphysics of power. All systems of avocation—conventional rituals, political manifestos or software standards—should be part of any critical examination of politics and ethics.

ENDNOTE

1. I used the term "avocation" to refer to individuated cultural resources which are systematically gathered to be invoked. These particularly include features on computers, but also encompass mnemonic phrasings in oral culture and clichés in print. This usage deviates somewhat from the standard use of the term, avocation, which refers to a person's hobbies or private passions, as opposed to their professional vocation.

REFERENCES

Austin, J. L. (1975) *How To Do Things With Words* Oxford: Clarendon Press.

Bulfinch, Thomas (1993) *The Golden Age of Myth and Legend* Hertfordshire: Wordsworth Editions.

Certeau, Michel de (1988) *The Practice of Everyday Life* Berkeley, Los Angeles and London: University of California Press.

Davis, Erik (1998) *Techgnosis. Myth, Magic and Mysticism in the Age of Information* New York: Harmony Books.

Derrida, Jacques (1976) *Of Grammatology* Baltimore, Maryland: The John Hopkins University Press.

Dery, Mark (1996) *Escape Velocity. Cyberculture at the End of the Century* London: Hodder & Stoughton.

Dreyfus Hubert (1966) "Alchemy and Artificial Intelligence" Rand paper P3244 Washington, DC: The Rand Corporation.

Fagles, Robert (transl.) and Homer (1996) *The Odyssey* New York: Viking Penguin.

Foley, John Miles (1995) *The Singer of Tales in Performance* Bloomington: Indiana University Press.

Foucault, Michel (1977) "What is an author?" in Donald F. Bouchard (ed.) *Language, Counter-memory, Practice* Ithaca: Cornell University Press 124–7.

Frazer, Sir James George (1960 [1922]) *The Golden Bough. A Study in Magic and Religion* London: MacMillan.

Heidegger, Martin (1977) "The question concerning technology" in Martin Heidegger (1977) *The Question Concerning Technology, and Other Essays* New York: Garland.

Heilbron, J. L. (1979) *Electricity in the 17th and 18th Centuries: A Study of Early Modern Physics* Berkeley: University of California Press.

Kardec, Allan (1874) *Experiental Spiritism: Book on Mediums, or, Guide for Mediums and Invocators* Wellingborough: Aquarian Press.

Latour, Bruno (1993) *We Have Never Been Modern* Cambridge, Mass.: Harvard University Press.

Lévi-Strauss, Claude and Felicity Baker (translator) (1987 [1950]) *Introduction to the Work of Marcel Mauss* [*Introduction à l'oeuvre de Marcel Mauss*] London: Routlege & Kegan Paul.

Mauss, Marcel (2001 [1902]) *A General Theory of Magic* London and Boston: Routledge Classics.

Melbourne Museum (2001) CSIRAC display *@digital.au* museum exhibit Melbourne: Australia.

Moravec, Hans (1988) *Mind Children—The Future of Robot and Human Intelligence* Cambridge, Mass.: Harvard University Press.

---------------- (1998) *Robot: Mere Machine to Transcendent Mind* Oxford: Oxford University Press.

Noble, David F. (1997) *The Religion of Technology: The Divinity of Man and the Spirit of Invention* New York: A. A. Knopf.

Ong, Walter (1982) *Orality and Literacy: the Technologizing of the Word* London and New York: Methuen.

Plato (1974) *The Republic* London: Penguin.

Rushkoff, Douglas (1994) *Cyberia. Life in the Trenches of Hyperspace* London: HarperCollins.

Schaffer, Simon and Steven Shapin (1985) *Leviathan and the Air-pump : Hobbes, Boyle, and the Experimental Life: including a translation of Thomas Hobbes,* Dialogus physicus de natura aeris Princeton, N.J.: Princeton University Press.

Seligman, Kurt (1997 [1947]) *The History of Magic* New York.

Thomas, Keith (1971) *Religion and the Decline of Magic* London: Weidenfeld & Nicolson.

Von Neumann, John (1958) *The Computer and the Brain* New Haven: Yale University Press.

Wark, McKenzie (1994) *Virtual Geography* Bloomington and Indianapolis: Indiana University Press.

Wertheim, Margaret (1999) *The Pearly Gates of Cyberspace* Sydney: Doubleday.

[EDWARD SCHEER]

Stelarc, magic and cyber-ritual

To be sure the "return of magic" may be just another story to while away the post-industrial night but it is precisely through such stories that technologies gain their character if not their lives (Erik Davis 1999: 188).

MAGIC AND RITUAL

In his manifesto *Parasite Visions*, Stelarc addresses the rather obvious negation of physicality experienced by internet users and proceeds, shamanistically, to evoke a brighter future for cybernetic corporeality:

As we hard-wire more high-fidelity image, sound, tactile and force-feedback sensation between bodies then we begin to generate powerful phantom presences—not phantom as in phantasmagorical, but phantom as in phantom limb sensation. The sensation of the remote body sucked onto your skin and nerve endings, collapsing the psychological and spatial distance between bodies on the Net. Just as in the experience of a Phantom Limb with the amputee, bodies will generate phantom partners, not because of a lack, but as extending and enhancing addition to their physiology. Your aura will not be your own (2000a: Part 10).

In this formulation, the discourse of the happy phantom alongside Stelarc's signature inventory of gear, constructs the familiar image of the ghost to comfort us in the age of the post-human and to suture the imago of high tech onto a troubled perception. Stelarc is using technology in this context to promote a kind of friendly ghost which is not so much hovering around reminding you of an absent dear one but helpfully suggesting absences you had not thought of, characters you might like to add to your image bank of dear ones, and perhaps also to enhance your repertoire of emotional responses to and/or through them. The image of the phantom limb here suggests not the sensation of the limb that was there and is now gone but the limbs that might be there but have not arrived yet. Your prosthesis is in the mail. In a sense Stelarc is saying that until we can acknowledge the intimate place of technology in our lives then we are all amputees, conceptually cut off from the physical reality of our own contemporary cybernetic lifestyles.

This passage is illustrative of the way Stelarc stages technology in his work, and gets it to perform conceptually. In order to indicate the way ahead he suggests we look way back to the origins of the technological age. Anticipating the future of intimate technology he abreacts the late nineteenth and early twentieth century with its auras and phantoms. For Stelarc, technology has a way of taking us back in time as well as forward in a kind of giff-like twostep. His performative play with extruded technology in events such as the huge hydraulic walking machine in *Exoskeleton*, the well-known third arm and more recent extended-arm events and even his *Movatar* action, recalls nineteenth-century Gothic visions of machines overpowering underperforming human peripherals. Similarly his visions of electronic phantoms recall the "monsters of Id" in *Forbidden Planet* (1956), nostalgic for troubled minds in which to dwell, and otherwise imperceptible with the power off. But like a latter-day Morpheus he calls us to examine our present and question our motives: "Bodies acting without expectation, producing movements without memory. Can a body act without emotion? Must a body continuously affirm its emotional, social and biological status quo?" (Stelarc http://www.stelarc.va.com.au) Motion without motivation? I will return to this question.

The visual excess and subtle hyperbole of his language is inseparable from his artwork with its excessive semio-tech display. Stelarc's sometimes geeky demonstrations are a dramatic mode of making visible a crisis in the

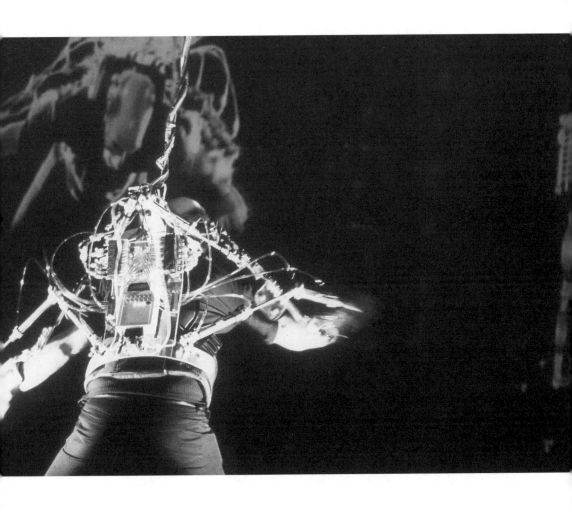

Stelarc performance of *Movatar* on Saturday 19 August 2000 at Casula Powerhouse, Sydney.
Photo: Heidrun Lohr, reproduced courtesy of the artist.

contemporary imaginary. Making visible overlooked ideas and experiences is what all artists do but constructing rituals in which to engage bodily with these materials is the domain of the performance artist. As I have argued elsewhere, performance art develops what Victor Turner calls "life crisis rituals" in response to drastically changing social and personal conditions. In Stelarc's performance art, it is the life crisis of an obsolete body finding itself without sympathetic environments in an age of technological innovation which is accelerating beyond the capacities of the organism to adapt. For him, it is a life crisis of the species. Performing it and inflecting it, making it livable, is the function of performance art for Stelarc, but his liminal cyber-rituals also imaginatively reinvent the crisis and project alternatives.

Ritual is the name of a practice often associated primarily with non-Western cultures because the West is so bad at them. We have inherited a miserable handful of etiolated rituals from our Christian forebears such as the confirmation rituals in Christian churches in which, in recent times, the young initiates are touched by the bishop rather than slapped as was once the case. The force of the ritual in terms of its physical effects is diminished. Nuptial and funerary rites continue to flourish which perhaps has to do with the continuous cultural hard-sell of hetero-romance on the one hand and the massive denial of death on the other. Death comes every day as a surprise to the West. The funeral is the only way we can deal with the scandal of the failure of our technologies to provide us with a fully functioning body after its own capacity for animation has gone. But in nuptial and confirmation rites what is it that unites a couple in the marriage ceremony, or locks the adolescent in to a Christian identity at a time when no identity seems possible? The power of ritual is still linked to authorised inscriptions of magical force—inexplicable, irrational, transformative and performative—onto contemporary life.

Consider the way that Michael Taussig has written about ritual, for instance, since his *Shamanism, Colonialism and the Wild Man* (1987). In the second half of this he discusses the healing rituals performed by the Indian shamans in the Putamayo Valley, Colombia. Taussig observes that "the white population that once tortured and killed Indians also depended on them for magical cures" (Eakin 2001). This double edge of culture, capitalism's own "negative capability" is what Taussig tries to perform in his writing. In *Defacement: Public Secrecy And The Labour Of The Negative* (1999) he notes that it is in ritual that "the symbolic character of the symbol erases itself." (137). In ritual, there is, for Taussig, a generalised transgression of the symbolic so that signs magically return to the forces that generated them. In this sense his is a profoundly performative reading of culture in which it is the "mysterious force" of ritual that must be understood as a precondition of learning the bases of cultural value in modern life. For him, only a criticism which deals with the force of the irrational can do justice to the truth (that is, the secret) of the social drama. An extended notion of ritual provides such an avenue for study since it is here that "the very notion…of a line dividing illusion from reality is itself really illusory" (1999: 114).

Taussig's writing on ritual is redolent of the language of Antonin Artaud's theatre of cruelty: its *poeisis* and its passion. But the real key to the connection is the thought of force, the prioritising of the force of the ritual over representation and semiology is essential to both Taussig's work—latterly in relation to defacement—and to Artaud's. To illustrate, in Artaud's post-surreal artworks, some 50 drawings on paper, he burns cigarette holes in the pages, breaks through the surface, defaces his own drawings, attacks the forms that he himself constructs, in order to dramatise the fact that what these forms always steal is force, that is, the force which brought the images to the surface and leaves no trace of itself.[1] These questions of energy and excess are, for Artaud as for Taussig, more than aesthetic or theatrical matters. They represent not a critique of representation but of the very dynamics of transformation from which representation proceeds, what Taussig calls "the hard core of culture" (1999: 184) or what we might call, in this context, magic.

Stelarc's body suspension events ("with insertions into the skin") of the 1970s and 1980s—25 separate performances in which his body was penetrated with shark hooks and suspended from a variety of mobile and immobile structures—were and are, as hardcore an illustration of transgressive ritual as you could wish for.[2] Like Artaud's artworks they refused the comforts of the uninterrupted surface. Stelarc broke through his own skin to uncover the body's potentials for spatial arrangement, for testing and dramatising gravity's force and for performing connections to technologies. These pieces featured awkward technologies such as cranes and scaffolding but his more recent and more self evidently technologically enhanced events may be less familiar in terms of a discourse on ritual.

Yet Erik Davis links ritual and magic precisely in terms of technology and the production of the new. He argues that ritual occurs through the enacted imagination of the unprecedented and even the impossible, quoting Sam Webster's definition of ritual as "the principal technology for programming the human organism". The body is coded and educated in fully immersive ritual environments which set the parameters for its behaviours. Davis explains that "ritual serves as a kind of virtual theatre that cultivates or 'programs' intentions and spiritual experiences in participants" (Davis 1999: 183–4). Stelarc's rituals may not invoke the sacred aspects of contemporary community life in the way ritual theories might suggest but they would be an example of what Taussig describes as the "labour of the negative" or the defacement through which the sacred appears despite its desire for secrecy, in all its problematic insubstantiality. Stelarc's rituals appear to be profane and post-human lab work but they are also challenging how we think of ourselves as animate beings and direct us to question ourselves as to what we can say of the force which animates us. Davis calls this "creative magic" which, as opposed to the "arresting magic of the social spectacle" (1999: 178–9) "stir(s) up new ways of seeing and being in a world striated with invisible grids of technocultural engineering" (1999: 179).

As Davis argues, the magician creatively manipulates the phantasms of the social: the "images, stories and desires" through which the world makes itself apparent. These phantasms, as much as the tired old human organism itself, are the stuff, the material of Stelarc's artwork. Descartes and Freud are among his most frequently invoked ghosts of an age when having a body was seemingly optional for the human *Geist*. Their magic trick of conjuring the body away from the discussion as to what was essential in the philosophy or psychology of human being is one Stelarc does not replicate. But he is not a figurative artist and his performances do not seek to describe the state of things. They are performatives, they engage with the world and seek to alter its dynamics. Appropriately, Davis' magical paradigm is, he says, based on a performative rather than a representational strategy: "language and ritual do not objectively delineate the world but help bring it into being" (1999: 174). The function of this type of magic, in its refusal of the terms in which the quotidian experiences itself, parallels the use of ritual in Stelarc as a testing of the world rather than its depiction.

Right: Stelarc performance of *Movatar* on Saturday 19 August 2000 at Casula Powerhouse, Sydney. Photo: Heidrun Lohr, reproduced courtesy of the artist.

CYBORG RITUALS—*MOVATAR*

The World Premiere Performance of *Movatar* on Saturday 19 August 2000 at Casula Powerhouse in Sydney's western suburbs featured a shiny metal jacket suspended from a cable (a backpack for the travelling cyborg?). As with all Stelarc's contraptions,[3] it is an elegantly and exactly finished piece of techno sculpture which he calls a "Motion Prosthesis" (MP). (He trained as a sculptor, after all, and not or not only as a prophet of the post-human.)

The system is a Servo-mechanism, a machine for staging feedback loops, in this case from the avatar to the body and back. Stelarc likens the Motion Prosthesis to the "muscles of the avatar" and the six pedals (that is, midi pedals) on the floor of the stage to the avatar's remote organs in the physical world. He explains the use of these organs as follows:

> The mutation mode is one pedal. If I like what the avatar is doing to me, I might allow it to persist or I might wait to see what happens, as there is a completely unpredictable change of posture strings occurring. But if I hit the "mutate" switch, I can then choose one of the six other switches to select rates of mutation, thus feeding back into the avatar's code and prompting changes in its behaviour (2000a).

In this way Stelarc is able to "contribute to a change of choreography".[4]

Stelarc's use of the notion of the "inverse motion capture system" suggests one productive mode of interpreting the event as performance art and, therefore, as ritual. It is neither a simple demonstration of hardware nor is it designed to generate visual effects for use in, say, a narrative cinema context. It is a conceptual construction enacted on and through the body of the artist and which is designed to make visible a cybernetic interaction (an organic/non-organic feedback system). He says that *Movatar* is "not a normal motion capture system in that it's not simply a physical body mapping its movements onto a computer model. Instead, you have feedback loops that are constantly modifying and modulating the relationship between the two" (2000a).[5] In *Movatar* the computer model maps its motion onto the physical body and so has to tread carefully. In theory, it could swivel its head through 360 degrees of rotation or bend its arms backwards which is why due caution was exercised in the programming phase of the project. This caution partly accounts for the severely limited range of movements exhibited by the avatar in its first performance.[6]

As Stelarc explains, this is also a function of the engineering of the MP in relation to the available bending motions of the human arms. "Ideally", he says, "as the avatar becomes more autonomous (i.e. develops more artificial intelligence), the relationship will become more complex. I looked at the avatar as a kind of viral life form, dormant in its computer state but able to generate an effect once it is connected to a host body" (2000a). As a virus, the avatar infects the body in barely visible ways. This bare visibility is one of the difficulties with the *Movatar* performance, as so many of the interesting interrelations are not communicable to the audience without the testimony of the artist. For the record, Stelarc says that "It was exciting to feel the modifying effects of the code and of the feedback from the real world" (2000a). While motion capture subtracts the body's appearance and leaves a record of its motion, inverse motion capture subtracts the body's motions, making it hostage to a (hopefully) benign virtual entity that just wants you for your body.[7]

Poor Stelarc. For years he has endured the misplaced shamanic references in critical studies of his work mostly in relation to his body suspensions.[8] For instance, Anne Marsh's reading of the suspensions in terms of "masochism", "shamanism", "sexual rites", "catharsis" and "ritual" returns the work to an order of subjective experience Stelarc himself repudiates (Marsh 2000: 178). He resists these readings, arguing that the subjective experience of his own body is not what his work is exploring and that such readings reduce his performances to experiments on the testing of his own personal limits. For him the shamanic reading or the modern primitivist reading detracts from the aesthetic and compositional aspects of the work and their cultural significance. The reading of his work proposed here in terms of a cyber-ritual is not an attempt to return it to modern primitivism or to the artists's own subjectivity or even the notion of their presence to guarantee the meaning of the performance as in the shamanism of the New Age movement, but instead it is directed at understanding what cultural work these performances actually do.

Stelarc's performances function as liminal rituals in that they make visible the passage of Western culture into a post-human relation to technology. They enable the perception that a properly cybernetic way of living and being in the world is already open to us. All his performances embody, in the enhanced symbolic behaviour characteristic of performance art, the myriad connectivities already operating between organic human and non-organic machinic entities. In

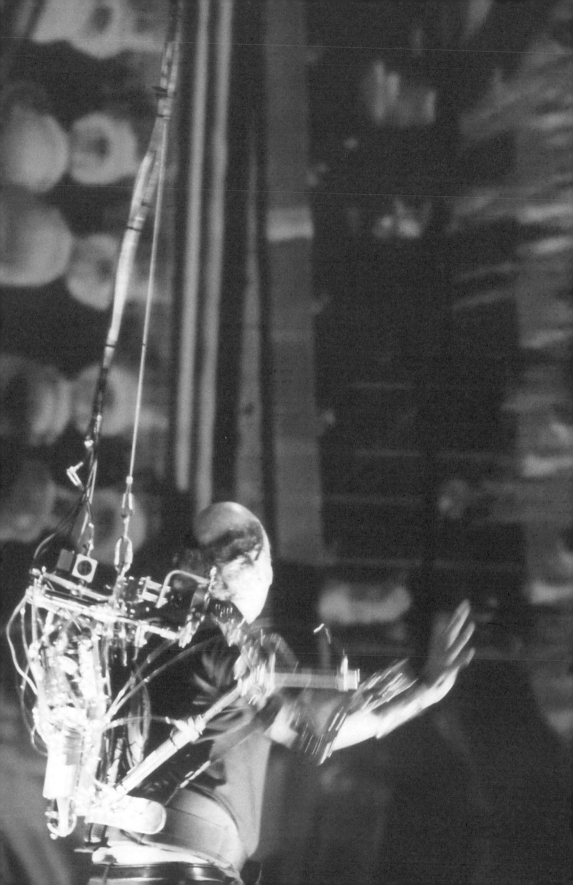

this respect Stelarc has consistently challenged the way our culture has imagined the body, whether it is seen as a sacred object, a fetish of the natural, an organic unity... and the culture has not always kept pace with him. While Stelarc is certainly of the generation of major artists who have used the body as the work of art itself (Marina Abramovic, Vito Acconci, Mike Parr, etc.) and manipulated it as an artefact rather than as a biological given (and therefore a kind of destiny), he is more concerned with the cybernetic body than with subjectivity, and more involved with pluralising and problematising the ways we speak of bodies and imagine them. He is also concerned with how we get them to do things and how they might do them differently. These are properly performative considerations but in the context of an event such as *Movatar* in which a human agent is not at the origin of the motivation for the action, it becomes more properly a question of sympathetic magic.

In *Mimesis and Alterity* Taussig identifies two primary modes of sympathetic magic in the work of George Frazer to do with similarity and contact. The first, similarity, relates to the principle of the copy, the effigy in which "the magician infers that he can produce any effect he desires merely by imitating it" and the second principle has to do with the idea that "things which have once been in contact with each other continue to act on each other at a distance after the physical contact has been severed" (Taussig 1993: 47). He quotes Freud's observation that "one of the most widespread magical procedures for damaging an enemy is to make an effigy of him" and that "[w]hether the effigy resembles him is of little account: any object can be 'made into' an effigy of him" (1993: 52).

Ultimately Taussig argues that the two modes of magic are necessarily interconnected since the copy or the image of the powerful magical object is usually problematically inaccurate, a "poorly executed ideogram" in the words of Mauss and Hubert's *A General Theory of Magic* (51) but still effective as a magical object because of its material connections ("those established by attaching hair, nail cuttings, pieces of clothing, and so forth to the likeness") (Taussig 1993: 57). Stelarc's rituals of physical augmentation observe similar magical structures of the "copy that is not a copy" taking effect due to its contiguity to the object of transformation: the body. Here it is the avatar, a crude animation sequence with

Left: Stelarc performance of *Movatar* on Saturday 19 August 2000 at Casula Powerhouse, Sydney. Photo: Heidrun Lohr, reproduced courtesy of the artist.

limited behaviours, disconnected but desiring intimacy. It "knows" that through the touch of a human agent and an actuator or motion prosthesis device, its kinetic potential, to mobilise a physical entity in the world is realised. Through Stelarc's performance this magic act by which an avatar choreographs and stimulates movement in an organic entity can take place in such a way that the object, this semi-autonomous virtual entity, forms a living relation with the world.

Stelarc's generosity in making his body available as the prosthesis for the virtual entity is unsurprising to those (humans and avatars) who know him. For without the embodiment provided thus far by his body, connected through the MP, *Movatar* is confined to a rather predictable existence in cyber space and a dull regime of jerky gestures. At least this way he gets to go out occasionally and see the world. Stelarc for his part operates not only as an organic substrate for a kinetic art project but as a kind of secular shaman,[9] channelling electricity rather than spirits, and evoking technological ghosts through rituals such as the *Movatar* performance.

Through calling up this digital double Stelarc radically questions our physical being in the world: are we zombies moving through the world with no autonomy directed by an external intelligence? Or are we cyborgs feeding back our own sense data into the machine world in which we operate? But it also continually reminds us that the loop of our sensory apparatus always takes us back into the world. As Susan Buck-Morss notes: "The circuit from sense perception to motor response begins and ends in the world" (Buck-Morss 1992:12). So the human element forms part of the circuit but is not the entire system itself. "We have always been prosthetic bodies, augmented and extended" as Stelarc asserts so the concept of cybernetic connection should not trouble us if we reflect on how extensively technological innovations have been adapted into our lives.

The magic Stelarc effects is perhaps just this shift in consciousness. By performing elaborate rituals of cybernetic connectivity he conjures an inchoate awareness of the extent to which we are already cyborgs. Our cars, mobile phones and hands-free extensions, wireless application protocols, and so on, continue collapsing distance into virtual proximity in a way which also pulls apart the absolute division between human and machine. But Stelarc also wants us to reflect on how these changes in our quotidian performance in the world themselves feed back into human/machine evolution. He emphasises that we need to interact more effectively and more empathetically with the increasingly denaturalised environment in which we, the other cyborgs, find ourselves. Accordingly, his event performances imagine the extension of the body's capacities for action in a way which suggest that the idea of a "natural body" disconnected from technology is itself a phantom, a ghost of the Enlightenment.

ENDNOTES

1. Graphics + Director/VRML Programming Gary Zebington; AI Director, Midi + Java Programming Damien Everett; Sound + Technical Design Rainer Linz; Website Movatar Animations Steve Middleton; Website Sound Andrew Garton. Thanks to SMC (Germany) and F18, respectively, for sponsoring and constructing *Movatar*'s motion prosthesis. Stelarc's presence here was part of the *Cyber Cultures: Sustained Release* program at Casula Powerhouse Arts Centre, in particular the "exhibition capsule" with the theme of *Posthuman Bodies* curated by Kathy Cleland.

2. Two iconic instances of this practice were the *Sitting/Swaying event for rock suspension* (Tamura Tokyo—11 May 1980) in which Stelarc was suspended in a sitting position encircled by 18 granite rocks, which coun-erbalanced his weight for a duration of approximately 17 minutes; and the *City Suspension* (above the Royal Theatre, Copenhagen—28 June 1985). In this latter piece Stelarc's body was connected to steel cables and hoisted through the air by an enormous industrial crane for a 24-minute period, achieving an altitude of 56 metres above street level. Stelarc was characteristically unclothed for these performances.

3. By now there are many: the *Third Arm*, which has recently been auctioned off after doing its duty these past 15 years, the *Extended Arm* and the *Exoskeleton* walking machine are some of the more recently used objects.

4. The other modes used in this prototype performance were: i) synchronous mode in which "the avatar is performing graphically and making my arms move at the *same* time"; ii) dialogue mode which uses a delay mechanism and in which "it might do something and get me to do it later"; and iii) a kind of "solo mode in which it does its thing and I do my thing, though there's a constant dialogue of physical gesture" (Stelarc Interview 2000a).

5. A more conventional approach, if that is ever the right expression with respect to this artist, to motion capture can be found in his 3D-avatar head project, The PROSTHETIC HEAD which was developed using a facial motion capture system to replicate Stelarc's own facial expressions. The head is a screen-based interactive system which uses what the website calls "an ultra-sound sensor system that alerts it of the user's presence, enabling it to initiate a conversation". The PROSTHETIC HEAD has been shown at New Territories, Glasgow 2003, The ICA, London 2003, InterAccess, Toronto 2003 at TRANSFIGURE, The Australian Centre for the Moving Image ACMI 8 December 2003 to 9 May 2004 and at Sherman Galleries in Sydney through May 2005.

6. Stelarc insists on its status as a work in progress and that is how I will try to read it here: in terms of the event itself and the planned development of the system.

7. The Artificial Intelligence laboratory at MIT are working on a similar problem with their Cog robot which is being trained in embodied sociality. They also have an "emotional—social learning" experiment underway with a disembodied or less-embodied robot called Kismet which is really just an inside-out electronic head but surprisingly cute. Claudia Dreifus, "Do Androids Dream? MIT Working on it" in *New York Times* "Science Times" section 7 November 2000 pF3.

8. In *Escape Velocity: Cyberculture at the end of the Twentieth Century* (1996) London: Hodder & Stoughton Dery accuses Stelarc of "narcissistic fantasies of complete isolation" (164). Anne Marsh's critique of the denial of gender in Stelarc's discourse of the body in *Body and Self* (1993) Melbourne: Oxford University Press is similarly caustic. See Jane Goodall "An Order of Pure Decision. Un-natural Selection in the work of Stelarc and Orlan" in *Body Modification* (2000) Mike Featherstone (ed.) London: Sage 149–84 for an excellent discussion of this issue.

9. Not to be confused with the shamanism of the modern primitives of the New Age.

REFERENCES

Buck-Morss, Susan (1992) "Aesthetics and Anaesthetics: Walter Benjamin's Essay Reconsidered" in 62 (October–Fall) 3–42.

Cacciari, Massimo (1996) *Posthumous People* (transl. Roger Friedman) Stanford: Stanford University Press.

Canny, John and Eric Paulos (2000) "Tele-Embodiment and Shattered Presence: Reconstructing The Body for Online Interaction" in Ken Goldberg (ed.) *The Robot in the Garden. Telerobotics and Telepistemology in the Age of the Internet* Cambridge, Mass.: MIT Press 276–95.

Carroll, David (1987) *Paraesthetics: Foucault Lyotard Derrida* New York and London: Methuen.

Davis, Erik (1999) *Techgnosis. Myth Magic and Mysticism in the Age of Information* London: Serpent's Tail.

Derrida, Jacques (1986) "Forcener le subjectile" in J. Derrida and Paule Thévenin *Antonin Artaud: Dessins et Portraits* Paris: Gallimard 55–109.

Dery, Mark (1996) *Escape Velocity: Cyberculture at the End of the Twentieth Century* London: Hodder & Stoughton.

Dreifus, Claudia "Do Androids Dream? MIT Working on it" in *New York Times* "Science Times" 7 November 2000 pF3.

Duchenne de Boulogne, Guillaume-Benjamin (1862) *Mécanisme de la Physiognomie Humaine* Paris: Jules Renouard.

Eakin, Emily "Anthropology's Alternative Radical" *New York Times* 21 April 2001.

Galison, Peter (1994) "The ontology of the enemy: Norbert Wiener and the Cybernetic Vision" *Critical Inquiry* 21 (Autumn) 228–66.

Goodall, Jane (2000) "An Order of Pure Decision. Un-natural Selection in the work of Stelarc and Orlan" in *Body Modification* Mike Featherstone (ed.) London: Sage 149–84.

Hillman, James (1979) *The Dream and the Underworld* New York: Harper & Row.

Marsh, Anne (1993) *Body and Self* Melbourne: Oxford University Press.

--------------"Obsolescent Bodies and Prosthetic Gods" (2000) in *Body Show/s: Australian Viewings of Live Performance* Peta Tait (ed.) Amsterdam and Atlanta: Rodopi 177–86.

Stelarc "Extra ear/Exoskeleton/Avatars" presentation at the Museum of Sydney 20 February 1999 (a dlux media/arts and Casula Powerhouse event).

---------- (2000a) Interview with Edward Scheer 26 August 2000 in Darlinghurst Sydney.

------------(2000b) *Parasite Visions* website HYPERLINK http://www.stelarc.va.com.au accessed 11 November 2000.

------------ (2000c) presentation at Performative Sites *Symposium Intersecting Art, technology, and the Body* 24 to 28 October at Penn State University, 25 October 2000.

------------2000 'Les Mutalogues—Avignonumerique' Performance–*Extended Arm* 5 May Grenier a Sel: Avignon France.

-----------1998 Kampnagel Performance–*Exoskeleton: Event For Walking Machine* 5 to 7 November, 11 to 13 November—Kampnagel: Hamburg, Germany.

-----------1982 *Handswriting* Maki Gallery Tokyo 22 May. Writing The Word 'Evolution' Simultaneously With Three Hands. Event for third hand.

Taussig, Michael (1993) *Mimesis and Alterity. A Particular History of the Senses* New York and London: Routledge.

------------(1999) *Defacement. Public Secrecy And The Labour Of The Negative* Stanford: Stanford University Press.

Turner, Victor (1986) *The Anthropology of Performance* New York: PAJ Publications.

[ACKNOWLEDGEMENTS]

The editors wish to thank Victoria Dawson and Roger Benjamin at Power Publications for their assistance and encouragement of this project. Julian Pefanis at Power facilitated this publication. We thank the authors for the care and quality of their pieces. We would also like to thank Stelarc for providing images of his work and Heidrun Löhr for the use of her photographs including the cover image. We acknowledge the financial support of the Department of Media and the Division of Society, Culture, Media and Philosophy at Macquarie University, Sydney and the Faculty of Arts and Social Sciences at the University of New South Wales, Sydney.

[CONTRIBUTORS]

CHRIS CHESHER is Director of the Arts Informatics Program at the University of Sydney. His work brings transdisciplinary media studies and cultural studies methods to bear on new media. He is actively involved with the Australasian critical new media and internet culture group *fibreculture*.

ANNE CRANNY-FRANCIS is Associate Professor of Critical and Cultural Studies at Macquarie University, Sydney. She has published widely on feminist fiction, media, cultural theory and literacy; her most recent work is on information technology, and includes the book *Multimedia: Texts and Contexts*.

ANNETTE HAMILTON is Dean of the Faculty of Arts and Social Sciences at the University of New South Wales, Sydney. She is an anthropologist who has been thinking about the relationships between cultural form, subjectivity, and imagination for many years. Her theoretical interests include psychoanalysis, screen cultures, and the impact of technological forms on cultural expression.

SCOTT McQUIRE is Senior Lecturer in the Media and Communications Program at the University of Melbourne. He is the author of *Crossing the Digital Threshold* (1997), *Visions of Modernity: Representation, Memory, Space and Time in the Age of the Camera* (1998), and *Maximum Vision* (1998). A forthcoming book is *The Uncanny Home* which explores interactions between media and urban space.

RACHEL MOORE lives in New York and London and writes on avant-garde film practice. She teaches at Goldsmiths College, University of London, in the Department of Media and Communications. She won the John Simon Guggenheim Fellowship for her forthcoming work, *In the Film Archive of Natural-History* on the use of degraded imagery in film. She is also the author of *Savage Theory, Cinema as Modern Magic* (2000).

STEPHEN MUECKE researches Cultural Studies at the University of Technology, Sydney. He is working on the historical and contemporary links between culture and commerce in the Indian Ocean. *Contingency in Madagascar* with photographer Max Pam, will be published by Les Imaginayres, Toulouse, France.

ANDREW MURPHIE is Senior Lecturer in the School of Media and Communications at the University of New South Wales, Sydney. He is co-author, with John Potts, of *Culture and Technology* (2003), and editor of the journal *fibreculture*. He has in the past worked as a freelance theatre director, and as a marketing manager and production manager for arts companies.

JOHN POTTS is Associate Professor in Media at Macquarie University, Sydney. He is the author of *Radio In Australia* (1989) and co-author with Andrew Murphie of *Culture and Technology* (2003). He has also published numerous journal articles and book chapters, on media arts and cyberculture, music and sound art, and intellectual history. He is a founding editor of *Scan*, the on-line journal of Media Arts and Culture.

PATRICIA PRINGLE is Senior Lecturer in Interior Design, the School of Architecture + Design at RMIT University, Melbourne. She is currently researching the ways in which modernity's new empathy with space, both imaginative and visceral, was manifested in popular amusements and entertainments of the late nineteenth and early twentieth centuries.

EDWARD SCHEER is Senior Lecturer in the School of Media, Film and Theatre at the University of New South Wales, Sydney. He is editor of *100 Years of Cruelty: Essays on Artaud* (2000) and *Antonin Artaud: A Critical Reader* (2003) and a founding editor of the journal *Performance Paradigm*. He is currently completing a study of the performance art of Mike Parr.

[INDEX]

3226